The Virtual Laboratory

Aristid Lindenmayer 1925–1989

Przemyslaw Prusinkiewicz
Aristid Lindenmayer

The Algorithmic Beauty
of Plants

With
James S. Hanan
F. David Fracchia
Deborah R. Fowler
Martin J. M. de Boer
Lynn Mercer

With 150 Illustrations, 48 in Color

Springer-Verlag
New York Berlin Heidelberg London Paris
Tokyo Hong Kong Barcelona Budapest

Przemyslaw Prusinkiewicz
Department of Computer Science
University of Regina
Regina, Saskatchewan S4S OA2
Canada

Series Editor
Przemyslaw Prusinkiewicz

Front jacket design: The roses in the foreground (*Roses* by D.R. Fowler, J. Hanan, and P. Prusinkiewicz [1990]) were modeled using L-systems. Distributed ray-tracing with one extended light source was used to simulate depth of field. The roses were placed on a background image (photography by G. Rossbach), which was scanned digitally and post-processed.

Back jacket design: *Water-lilies* by D.R. Fowler, J. Hanan, P. Prusinkiewicz, and N. Fuller (1990). A scene inspired by *Water-lilies Pool—Harmony in Green* by Claude Monet (1899). The water plants and the trees were modeled using L-systems, while the bridge was created using Euclidean geometry construction. The entire scene was ray-traced, then the resulting image was post-processed to create the appearance of an impressionistic painting.

Printed on acid-free paper.

Typeset by the authors using the LATeX document preparation system.
Printed and bound by: Kingsport Press, Kingsport, Tennessee.
Printed in the United States of America.

9 8 7 6 5 4 3 2

ISBN 0-387-97297-8 Springer-Verlag New York Berlin Heidelberg
ISBN 3-540-97297-8 Springer-Verlag Berlin Heidelberg New York

Preface

The beauty of plants has attracted the attention of mathematicians for centuries. Conspicuous geometric features such as the bilateral symmetry of leaves, the rotational symmetry of flowers, and the helical arrangements of scales in pine cones have been studied most extensively. This focus is reflected in a quotation from Weyl [159, page 3], "Beauty is bound up with symmetry."

Mathematics and beauty

This book explores two other factors that organize plant structures and therefore contribute to their beauty. The first is the elegance and relative simplicity of *developmental algorithms*, that is, the rules which describe plant development in time. The second is *self-similarity*, characterized by Mandelbrot [95, page 34] as follows:

> When each piece of a shape is geometrically similar to the whole, both the shape and the cascade that generate it are called self-similar.

This corresponds with the biological phenomenon described by Herman, Lindenmayer and Rozenberg [61]:

> In many growth processes of living organisms, especially of plants, regularly repeated appearances of certain multicellular structures are readily noticeable.... In the case of a compound leaf, for instance, some of the lobes (or leaflets), which are parts of a leaf at an advanced stage, have the same shape as the whole leaf has at an earlier stage.

Thus, self-similarity in plants is a result of developmental processes. By emphasizing the relationship between growth and form, this book follows a long tradition in biology. D'Arcy Thompson [143] traces its origins to the late seventeenth century, and comments:

Growth and form

> Organic form itself is found, mathematically speaking, to be a function of time.... We might call the form of an organism an *event in space-time*, and not merely a *configuration in space.*

This concept is echoed by Hallé, Oldeman and Tomlinson [58]:

> The idea of the form implicitly contains also the history of such a form.

Modeling of plants

The developmental processes are captured using the formalism of *L-systems*. They were introduced in 1968 by Lindenmayer [82] as a theoretical framework for studying the development of simple multicellular organisms, and subsequently applied to investigate higher plants and plant organs. After the incorporation of geometric features, plant models expressed using L-systems became detailed enough to allow the use of computer graphics for realistic visualization of plant structures and developmental processes.

The emphasis on graphics has several motivations. A visual comparison of models with real structures is an important component of model validation. The display of parameters and processes not observable directly in living organisms may assist in the analysis of their physiology, and thus present a valuable tool for educational purposes. From an aesthetic perspective, plants present a wealth of magnificent objects for image synthesis. The quest for photorealism challenges modeling and rendering algorithms, while a departure from realism may offer a fresh view of known structures.

The application of computer graphics to biological structures is only one of many factors that contribute to the interdisciplinary character of this book. For example, the notion of L-systems is a part of formal language theory, rooted in the theory of algorithms. The application of L-systems to plant description has been studied by biologists, and involves various methods of general mathematics. Self-similarity relates plant structures to the geometry of fractals. Computer-aided visualization of these structures, and the processes that create them, joins science with art.

About the book

The study of an area that combines so many disciplines is very stimulating. Some results may be of special interest to students of biology or computer graphics, but a much wider circle of readers, generally interested in science, may find mathematical plant models inspiring, and the open problems worth further thought. Consequently, all basic concepts are presented in a self-contained manner, assuming only general knowledge of mathematics at the junior college level.

This book focuses on original research results obtained by the authors in the scope of the cooperation between the Theoretical Biology Group, directed by Aristid Lindenmayer at the University of Utrecht, and the Computer Graphics Group, working under the supervision of Przemyslaw Prusinkiewicz at the University of Regina. Technically, the book evolved from the SIGGRAPH '88 and '89 course notes *Lindenmayer systems, fractals, and plants*, published by Springer-Verlag in the series *Lecture Notes in Biomathematics* [112]. The present volume has been extended with edited versions of recent journal and conference papers (see Sources), as well as previously unpublished results.

Aristid Lindenmayer is the author of the notion of L-systems which forms the main thread of the book. He also played an essential role in the reported research by suggesting topics for study, guiding the construction of specific plant models, monitoring their correctness and participating in many discussions of biological and mathematical problems. Seriously ill, Professor Lindenmayer co-authored and edited several chapters, but was not able to participate in the completion of this work. If any inaccuracies or mistakes remain, he could not prevent them. Still, in spite of unavoidable shortcomings, we hope that this book will convey his and our excitement of applying mathematics to explore the beauty of plants.

Acknowledgements

While preparing this book, we received extraordinary support and help from many people, and we are deeply thankful to all of them. First of all, we would like to thank those who were directly involved in the underlying research and software development. Craig Kolb wrote the ray tracer *rayshade* used to render many of the images included in the book. Allan Snider developed several software tools, including a previewer for *rayshade*, and provided valuable expertise in ray-tracing. Daryl Hepting developed software for rendering sets defined by iterated function systems and provided diagrams for Chapter 8. Norma Fuller modeled several man-made objects incorporated into the images.

We would like to thank Zdzisław Pawlak and Grzegorz Rozenberg who initiated the contact between the Theoretical Biology Group at the University of Utrecht and the Computer Graphics Group at the University of Regina. Benoit Mandelbrot and Heinz-Otto Peitgen made it possible to conduct parts of the reported research at Yale University and the University of Bremen.

We are also grateful to all those who shared their knowledge with us and made suggestions reflected in this book. Discussions and correspondence with Jules Bloomenthal, Mark de Does, Pauline Hogeweg, Jacqueline and Hermann Lück, Gavin Miller, Laurie Reuter, Dietmar Saupe and Alvy Ray Smith were particularly fruitful.

Research reported in this book was funded by grants from the Natural Sciences and Engineering Research Council of Canada, as well as an equipment donation and a research grant from Apple Computer, Inc. We are particularly grateful to Mark Cutter for making the support from Apple possible. The Graphics Laboratory at the University of Regina also enjoys continued support from the university. The influence of Lawrence Symes and R. Brien Maguire is deeply appreciated. In addition, the University of Regina and the University of Utrecht contributed towards travel expenses.

We would like to thank Springer-Verlag and in particular, Gerhard Rossbach and Nina LaVoy from the Springer West Coast Office, for the expedient publishing of this book.

Finally, we would like to thank our families and friends for their love, support and patience while we worked on this book.

Przemyslaw Prusinkiewicz
James Hanan
F. David Fracchia
Deborah R. Fowler
Martin J. M. de Boer
Lynn Mercer
Regina, Canada
May 1990

Sources

In the preparation of this book, edited parts of the following publications were used:

- P. Prusinkiewicz and J. Hanan. *Lindenmayer systems, fractals, and plants*, volume 79 of *Lecture Notes in Biomathematics*. Springer-Verlag, Berlin, 1989. [Chapters 1, 3, 5]

- P. Prusinkiewicz, A. Lindenmayer and J. Hanan. Developmental models of herbaceous plants for computer imagery purposes. Proceedings of SIGGRAPH '88 (Atlanta, Georgia, August 1-5, 1988), in *Computer Graphics*, 22(4):141–150, 1988. Used with the permission of the Association for Computing Machinery. [Chapters 1, 3, 5]

- P. Prusinkiewicz and J. Hanan. Visualization of botanical structures and processes using parametric L-systems. In D. Thalmann, editor, *Scientific visualization and graphics simulation*, pages 183–201. J. Wiley & Sons, 1990. [Chapters 1–5]

- D. Fowler, J. Hanan and P. Prusinkiewicz. Modelling spiral phyllotaxis. *computers & graphics*, 13(3):291–296, 1989. [Chapter 4]

- J. S. Hanan. PLANTWORKS: A software system for realistic plant modelling. Master's thesis, University of Regina, 1988. [Chapter 5]

- F. D. Fracchia, P. Prusinkiewicz and M. J. M. de Boer. Animation of the development of multicellular structures. In N. Magnenat-Thalmann and D. Thalmann, editors, *Computer Animation '90*, pages 3–18. Springer-Verlag, Tokyo, 1990. [Chapter 7]

- F. D. Fracchia, P. Prusinkiewicz, and M. J. M. de Boer. Visualization of the development of multicellular structures. In *Proceedings of Graphics Interface '90*, pages 267–277, 1990. [Chapter 7]

- L. Mercer, P. Prusinkiewicz, and J. Hanan. The concept and design of a virtual laboratory. In *Proceedings of Graphics Interface '90*, pages 149–155, 1990. [Appendix A]

Contents

Chapter 1

Graphical modeling using L-systems

Lindenmayer systems — or L-systems for short — were conceived as a mathematical theory of plant development [82]. Originally, they did not include enough detail to allow for comprehensive modeling of higher plants. The emphasis was on plant topology, that is, the neighborhood relations between cells or larger plant modules. Their geometric aspects were beyond the scope of the theory. Subsequently, several geometric interpretations of L-systems were proposed with a view to turning them into a versatile tool for plant modeling. Throughout this book, an interpretation based on turtle geometry is used [109]. Basic notions related to L-system theory and their turtle interpretation are presented below.

1.1 Rewriting systems

The central concept of L-systems is that of rewriting. In general, rewriting is a technique for defining complex objects by successively replacing parts of a simple initial object using a set of *rewriting rules* or *productions*. The classic example of a graphical object defined in terms of rewriting rules is the *snowflake curve* (Figure 1.1), proposed in 1905 by von Koch [155]. Mandelbrot [95, page 39] restates this construction as follows:

Koch construction

> One begins with *two shapes*, an *initiator* and a *generator*. The latter is an oriented broken line made up of N equal sides of length r. Thus each stage of the construction begins with a broken line and consists in replacing each straight interval with a copy of the generator, reduced and displaced so as to have the same end points as those of the interval being replaced.

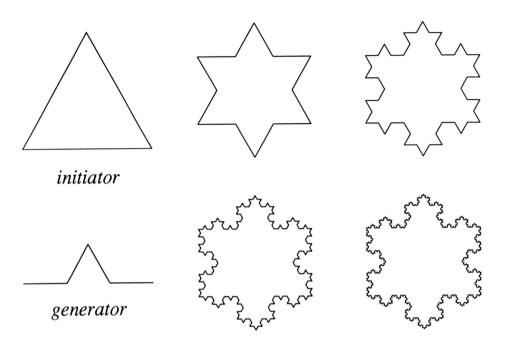

Figure 1.1: Construction of the snowflake curve

While the Koch construction recursively replaces open polygons, rewriting systems that operate on other objects have also been investigated. For example, Wolfram [160, 161] studied patterns generated by rewriting elements of rectangular arrays. A similar array-rewriting mechanism is the cornerstone of Conway's popular *game of life* [49, 50]. An important body of research has been devoted to various graph-rewriting systems [14, 33, 34].

Grammars   The most extensively studied and the best understood rewriting systems operate on character strings. The first formal definition of such a system was given at the beginning of this century by Thue [128], but a wide interest in string rewriting was spawned in the late 1950s by Chomsky's work on formal grammars [13]. He applied the concept of rewriting to describe the syntactic features of natural languages. A few years later Backus and Naur introduced a rewriting-based notation in order to provide a formal definition of the programming language ALGOL-60 [5, 103]. The equivalence of the Backus-Naur form (BNF) and the context-free class of Chomsky grammars was soon recognized [52], and a period of fascination with syntax, grammars and their application to computer science began. At the center of attention were sets of strings — called formal languages — and the methods for generating, recognizing and transforming them.

L-systems   In 1968 a biologist, Aristid Lindenmayer, introduced a new type of string-rewriting mechanism, subsequently termed L-systems [82]. The essential difference between Chomsky grammars and L-systems lies in

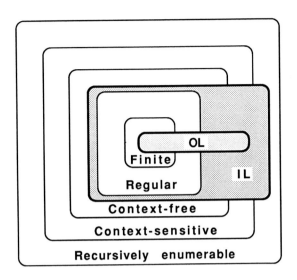

Figure 1.2: Relations between Chomsky classes of languages and language classes generated by L-systems. The symbols OL and IL denote language classes generated by context-free and context-sensitive L-systems, respectively.

the method of applying productions. In Chomsky grammars productions are applied sequentially, whereas in L-systems they are applied in parallel and simultaneously replace all letters in a given word. This difference reflects the biological motivation of L-systems. Productions are intended to capture cell divisions in multicellular organisms, where many divisions may occur at the same time. Parallel production application has an essential impact on the formal properties of rewriting systems. For example, there are languages which can be generated by context-free L-systems (called OL-systems) but not by context-free Chomsky grammars [62, 128] (Figure 1.2).

1.2   DOL-systems

This section presents the simplest class of L-systems, those which are deterministic and context-free, called DOL-systems. The discussion starts with an example that introduces the main idea in intuitive terms.

Consider strings (words) built of two letters a and b, which may *Example* occur many times in a string. Each letter is associated with a rewriting rule. The rule $a \rightarrow ab$ means that the letter a is to be replaced by the string ab, and the rule $b \rightarrow a$ means that the letter b is to be replaced by a. The rewriting process starts from a distinguished string called the axiom. Assume that it consists of a single letter b. In the first derivation step (the first step of rewriting) the axiom b is replaced

Figure 1.3: Example of a derivation in a DOL-system

by a using production $b \rightarrow a$. In the second step a is replaced by ab using production $a \rightarrow ab$. The word ab consists of two letters, both of which are *simultaneously* replaced in the next derivation step. Thus, a is replaced by ab, b is replaced by a, and the string aba results. In a similar way, the string aba yields $abaab$ which in turn yields $abaababa$, then $abaababaabaab$, and so on (Figure 1.3).

Formal definitions describing DOL-systems and their operation are given below. For more details see [62, 127].

L-system   Let V denote an alphabet, V^* the set of all words over V, and V^+ the set of all nonempty words over V. A *string OL-system* is an ordered triplet $G = \langle V, \omega, P \rangle$ where V is the *alphabet* of the system, $\omega \in V^+$ is a nonempty word called the *axiom* and $P \subset V \times V^*$ is a finite *set of productions*. A production $(a, \chi) \in P$ is written as $a \rightarrow \chi$. The letter a and the word χ are called the *predecessor* and the *successor* of this production, respectively. It is assumed that for any letter $a \in V$, there is at least one word $\chi \in V^*$ such that $a \rightarrow \chi$. If no production is explicitly specified for a given predecessor $a \in V$, the *identity production* $a \rightarrow a$ is assumed to belong to the set of productions P. An OL-system is *deterministic* (noted *DOL-system*) if and only if for each $a \in V$ there is exactly one $\chi \in V^*$ such that $a \rightarrow \chi$.

Derivation   Let $\mu = a_1 \ldots a_m$ be an arbitrary word over V. The word $\nu = \chi_1 \ldots \chi_m \in V^*$ is *directly derived* from (or *generated* by) μ, noted $\mu \Rightarrow \nu$, if and only if $a_i \rightarrow \chi_i$ for all $i = 1, \ldots, m$. A word ν is generated by G in a derivation of *length* n if there exists a *developmental sequence* of words $\mu_0, \mu_1, \ldots, \mu_n$ such that $\mu_0 = \omega$, $\mu_n = \nu$ and $\mu_0 \Rightarrow \mu_1 \Rightarrow \ldots \Rightarrow \mu_n$.

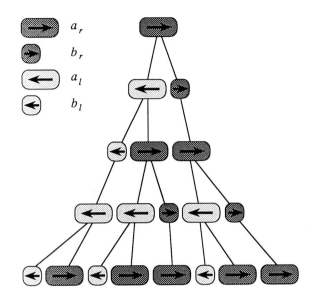

Figure 1.4: Development of a filament (*Anabaena catenula*) simulated using a DOL-system

The following example provides another illustration of the operation of DOL-systems. The formalism is used to simulate the development of a fragment of a multicellular filament such as that found in the blue-green bacteria *Anabaena catenula* and various algae [25, 84, 99]. The symbols a and b represent cytological states of the cells (their size and readiness to divide). The subscripts l and r indicate cell polarity, specifying the positions in which daughter cells of type a and b will be produced. The development is described by the following L-system:

Anabaena

$$
\begin{array}{rl}
\omega : & a_r \\
p_1 : & a_r \rightarrow a_l b_r \\
p_2 : & a_l \rightarrow b_l a_r \\
p_3 : & b_r \rightarrow a_r \\
p_4 : & b_l \rightarrow a_l
\end{array}
\qquad (1.1)
$$

Starting from a single cell a_r (the axiom), the following sequence of words is generated:

$$
\begin{array}{l}
a_r \\
a_l b_r \\
b_l a_r a_r \\
a_l a_l b_r a_l b_r \\
b_l a_r b_l a_r a_r b_l a_r a_r \\
\cdots
\end{array}
$$

Under a microscope, the filaments appear as a sequence of cylinders of various lengths, with a-type cells longer than b-type cells. The corresponding schematic image of filament development is shown in Figure 1.4. Note that due to the discrete nature of L-systems, the continuous growth of cells between subdivisions is not captured by this model.

1.3   Turtle interpretation of strings

The geometric interpretation of strings applied to generate schematic images of *Anabaena catenula* is a very simple one. Letters of the L-system alphabet are represented graphically as shorter or longer rectangles with rounded corners. The generated structures are one-dimensional chains of rectangles, reflecting the sequence of symbols in the corresponding strings.

Previous methods

In order to model higher plants, a more sophisticated graphical interpretation of L-systems is needed. The first results in this direction were published in 1974 by Frijters and Lindenmayer [46], and Hogeweg and Hesper [64]. In both cases, L-systems were used primarily to determine the branching topology of the modeled plants. The geometric aspects, such as the lengths of line segments and the angle values, were added in a post-processing phase. The results of Hogeweg and Hesper were subsequently extended by Smith [136, 137], who demonstrated the potential of L-systems for realistic image synthesis.

Szilard and Quinton [141] proposed a different approach to L-system interpretation in 1979. They concentrated on image representations with rigorously defined geometry, such as chain coding [43], and showed that strikingly simple DOL-systems could generate the intriguing, convoluted curves known today as *fractals* [95]. These results were subsequently extended in several directions. Siromoney and Subramanian [135] specified L-systems which generate classic space-filling curves. Dekking investigated the limit properties of curves generated by L-systems [32] and concentrated on the problem of determining the fractal (Hausdorff) dimension of the limit set [31]. Prusinkiewicz focused on an interpretation based on a LOGO-style turtle [1] and presented more examples of fractals and plant-like structures modeled using L-systems [109, 111]. Further applications of L-systems with turtle interpretation include realistic modeling of herbaceous plants [117], description of *kolam* patterns (an art form from Southern India) [112, 115, 133, 134], synthesis of musical scores [110] and automatic generation of space-filling curves [116].

Turtle

The basic idea of turtle interpretation is given below. A *state* of the turtle is defined as a triplet (x, y, α), where the Cartesian coordinates (x, y) represent the turtle's *position*, and the angle α, called the *heading*, is interpreted as the direction in which the turtle is facing. Given the *step size* d and the *angle increment* δ, the turtle can respond to

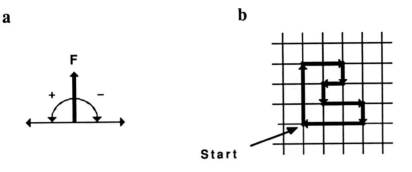

FFF-FF-F-F+F+FF-F-FFF

Figure 1.5: (a) Turtle interpretation of string symbols F, $+$, $-$. (b) Interpretation of a string. The angle increment δ is equal to $90°$. Initially the turtle faces up.

commands represented by the following symbols (Figure 1.5a):

F   Move forward a step of length d. The state of the turtle changes to (x', y', α), where $x' = x + d\cos\alpha$ and $y' = y + d\sin\alpha$. A line segment between points (x, y) and (x', y') is drawn.

f   Move forward a step of length d without drawing a line.

$+$ Turn left by angle δ. The next state of the turtle is $(x, y, \alpha+\delta)$. The positive orientation of angles is counterclockwise.

$-$ Turn right by angle δ. The next state of the turtle is $(x, y, \alpha - \delta)$.

Interpretation

Given a string ν, the initial state of the turtle (x_0, y_0, α_0) and fixed parameters d and δ, the *turtle interpretation* of ν is the figure (set of lines) drawn by the turtle in response to the string ν (Figure 1.5b). Specifically, this method can be applied to interpret strings which are generated by L-systems. For example, Figure 1.6 presents four approximations of the *quadratic Koch island* taken from Mandelbrot's book [95, page 51]. These figures were obtained by interpreting strings generated by the following L-system:

$$\omega : \quad F - F - F - F$$
$$p : \quad F \rightarrow F - F + F + FF - F - F + F$$

The images correspond to the strings obtained in derivations of length 0 to 3. The angle increment δ is equal to $90°$. The step length d is decreased four times between subsequent images, making the distance

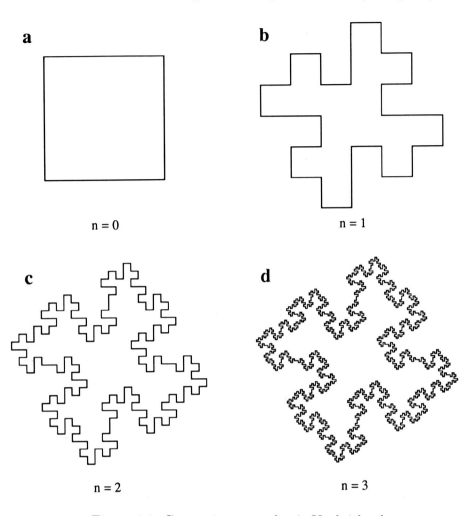

Figure 1.6: Generating a quadratic Koch island

*Koch
constructions
vs. L-systems*

between the endpoints of the successor polygon equal to the length of the predecessor segment.

The above example reveals a close relationship between Koch constructions and L-systems. The initiator corresponds to the axiom and the generator corresponds to the production successor. The predecessor F represents a single edge. L-systems specified in this way can be perceived as *codings* for Koch constructions. Figure 1.7 presents further examples of Koch curves generated using L-systems. A slight complication occurs if the curve is not connected; a second production (with the predecessor f) is then required to keep components the proper distance from each other (Figure 1.8). The ease of modifying L-systems makes them suitable for developing new Koch curves. For example, one can start from a particular L-system and observe the results of inserting, deleting or replacing some symbols. A variety of curves obtained this way are shown in Figure 1.9.

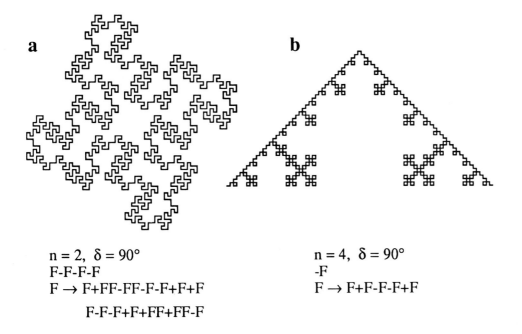

<table>
<tr><td>

a

n = 2, δ = 90°
F-F-F-F
F → F+FF-FF-F-F+F+F

F-F-F+F+FF+FF-F
</td><td>

b

n = 4, δ = 90°
-F
F → F+F-F-F+F
</td></tr>
</table>

Figure 1.7: Examples of Koch curves generated using L-systems: (a) Quadratic Koch island [95, page 52], (b) A quadratic modification of the snowflake curve [95, page 139]

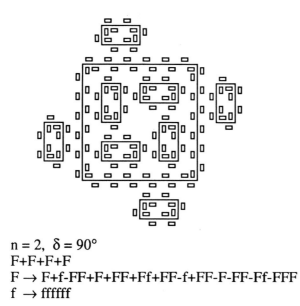

n = 2, δ = 90°
F+F+F+F
F → F+f-FF+F+FF+Ff+FF-f+FF-F-FF-Ff-FFF
f → ffffff

Figure 1.8: Combination of islands and lakes [95, page 121]

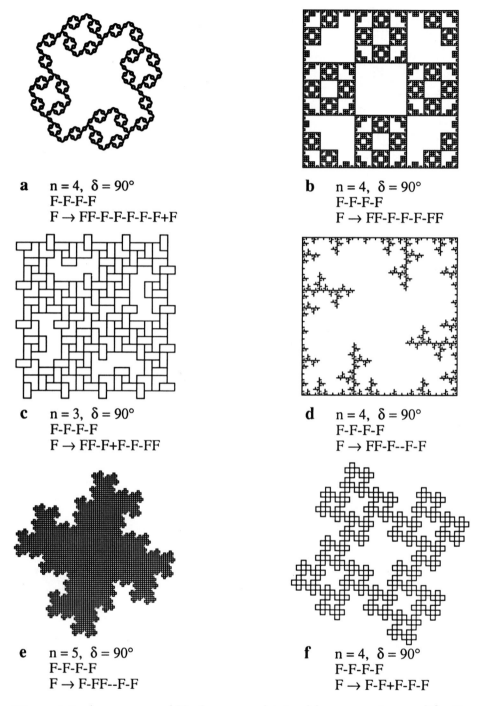

a   n = 4, δ = 90°
F-F-F-F
F → FF-F-F-F-F-F+F

b   n = 4, δ = 90°
F-F-F-F
F → FF-F-F-F-FF

c   n = 3, δ = 90°
F-F-F-F
F → FF-F+F-F-FF

d   n = 4, δ = 90°
F-F-F-F
F → FF-F--F-F

e   n = 5, δ = 90°
F-F-F-F
F → F-FF--F-F

f   n = 4, δ = 90°
F-F-F-F
F → F-F+F-F-F

Figure 1.9: A sequence of Koch curves obtained by successive modification of the production successor

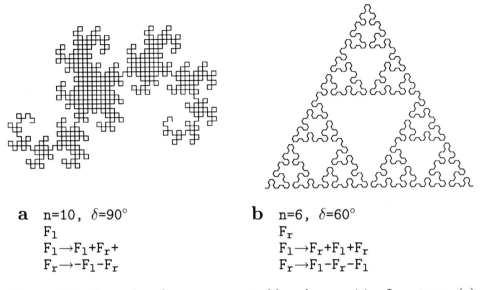

a n=10, δ=90°
F_l
$F_l \rightarrow F_l + F_r +$
$F_r \rightarrow -F_l - F_r$

b n=6, δ=60°
F_r
$F_l \rightarrow F_r + F_l + F_r$
$F_r \rightarrow F_l - F_r - F_l$

Figure 1.10: Examples of curves generated by edge-rewriting L-systems: (a) the dragon curve [48], (b) the Sierpiński gasket [132]

1.4 Synthesis of DOL-systems

Random modification of productions gives little insight into the relationship between L-systems and the figures they generate. However, we often wish to construct an L-system which captures a given structure or sequence of structures representing a developmental process. This is called the *inference problem* in the theory of L-systems. Although some algorithms for solving it were reported in the literature [79, 88, 89], they are still too limited to be of practical value in the modeling of higher plants. Consequently, the methods introduced below are more intuitive in nature. They exploit two modes of operation for L-systems with turtle interpretation, called *edge rewriting* and *node rewriting* using terminology borrowed from graph grammars [56, 57, 87]. In the case of edge rewriting, productions substitute figures for polygon edges, while in node rewriting, productions operate on polygon vertices. Both approaches rely on capturing the recursive structure of figures and relating it to a tiling of a plane. Although the concepts are illustrated using abstract curves, they apply to branching structures found in plants as well.

1.4.1 Edge rewriting

Edge rewriting can be viewed as an extension of Koch constructions. For example, Figure 1.10a shows the *dragon curve* [21, 48, 95] and the L-system that generated it. Both the F_l and F_r symbols represent edges created by the turtle executing the "move forward" command. The productions substitute F_l or F_r edges by pairs of lines forming

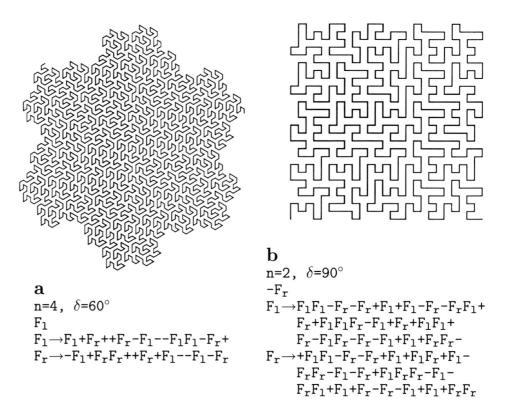

a

n=4, δ=60°

F_l

$F_l \rightarrow F_l + F_r + + F_r - F_l - - F_l F_l - F_r +$
$F_r \rightarrow -F_l + F_r F_r + + F_r + F_l - - F_l - F_r$

b

n=2, δ=90°

$-F_r$

$F_l \rightarrow F_l F_l - F_r - F_r + F_l + F_l - F_r - F_r F_l +$
$F_r + F_l F_l F_r - F_l + F_r + F_l F_l +$
$F_r - F_l F_r - F_r - F_l + F_l + F_r F_r -$
$F_r \rightarrow + F_l F_l - F_r - F_r + F_l + F_l F_r + F_l -$
$F_r F_r - F_l - F_r + F_l F_r F_r - F_l -$
$F_r F_l + F_l + F_r - F_r - F_l + F_l + F_r F_r$

Figure 1.11: Examples of FASS curves generated by edge-rewriting L-systems: (a) hexagonal Gosper curve [51], (b) quadratic Gosper curve [32] or E-curve [96]

FASS curve construction

left or right turns. Many interesting curves can be obtained assuming two types of edges, "left" and "right." Figures 1.10b and 1.11 present additional examples.

The curves included in Figure 1.11 belong to the class of *FASS* curves (an acronym for space-<u>f</u>illing, self-<u>a</u>voiding, <u>s</u>imple and self-<u>s</u>imilar) [116], which can be thought of as finite, self-avoiding approximations of curves that pass through *all* points of a square (space-filling curves [106]). McKenna [96] presented an algorithm for constructing FASS curves using edge replacement. It exploits the relationship between such a curve and a recursive subdivision of a square into tiles. For example, Figure 1.12 shows the tiling that corresponds to the E-curve of Figure 1.11b. The polygon replacing an edge F_l (Figure 1.12a) approximately fills the square on the left side of F_l (b). Similarly, the polygon replacing an edge F_r (c) approximately fills the square on the right side of that edge (d). Consequently, in the next derivation step, each of the 25 tiles associated with the curves (b) or (d) will be covered by their reduced copies (Figure 1.11b). A recursive application of this argument indicates that the whole curve is approximately space-filling. It is also self-avoiding due to the following two properties:

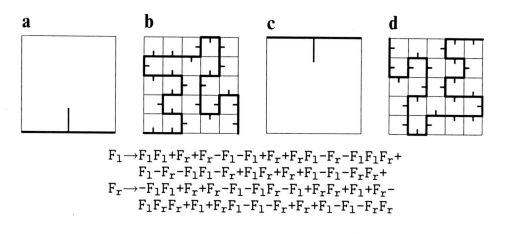

$$F_1 \rightarrow F_1F_1+F_r+F_r-F_1-F_1+F_r+F_rF_1-F_r-F_1F_1F_r+$$
$$F_1-F_r-F_1F_1-F_r+F_1F_r+F_r+F_1-F_1-F_rF_r+$$
$$F_r \rightarrow -F_1F_1+F_r+F_r-F_1-F_1F_r-F_1+F_rF_r+F_1+F_r-$$
$$F_1F_rF_r+F_1+F_rF_1-F_1-F_r+F_r+F_1-F_1-F_rF_r$$

Figure 1.12: Construction of the E-curve on the square grid. Left and right edges are distinguished by the direction of ticks.

- the generating polygon is self-avoiding, and

- no matter what the relative orientation of the polygons lying on two adjacent tiles, their union is a self-avoiding curve.

The first property is obvious, while the second can be verified by considering all possible relative positions of a pair of adjacent tiles.

Using a computer program to search the space of generating polygons, McKenna found that the E-curve is the simplest FASS curve obtained by edge replacement in a square grid. Other curves require generators with more edges (Figure 1.13). The relationship between edge rewriting and tiling of the plane extends to branching structures, providing a method for constructing and analyzing L-systems which operate according to the edge-rewriting paradigm (see Section 1.10.3).

1.4.2 Node rewriting

The idea of node rewriting is to substitute new polygons for nodes of the predecessor curve. In order to make this possible, turtle interpretation is extended by symbols which represent arbitrary subfigures. As shown in Figure 1.14, each subfigure A from a set of subfigures $\mathcal{A}$ is represented by:

Subfigures

- two *contact points*, called the *entry point* P_A and the *exit point* Q_A, and

- two *direction vectors*, called the *entry vector* $\vec{p}_A$ and the *exit vector* $\vec{q}_A$.

During turtle interpretation of a string ν, a symbol $A \in \mathcal{A}$ incorporates the corresponding subfigure into a picture. To this end, A is translated

a   **b**

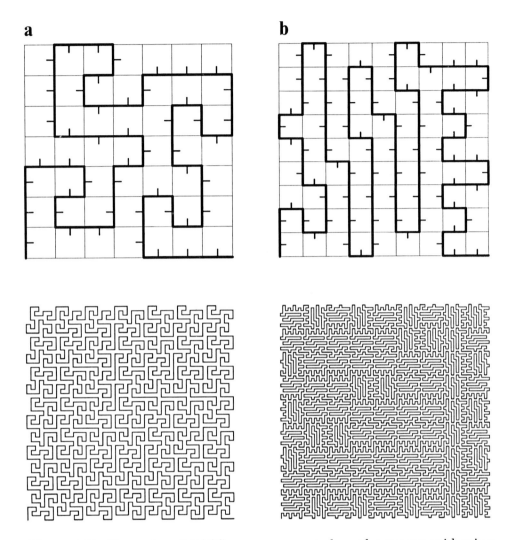

Figure 1.13: Examples of FASS curves generated on the square grid using edge replacement: (a) a SquaRecurve (grid size 7×7), (b) an E-tour (grid size 9×9). Both curves are from [96].

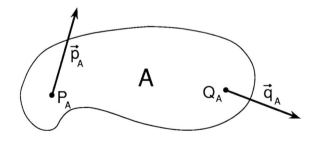

Figure 1.14: Description of a subfigure A

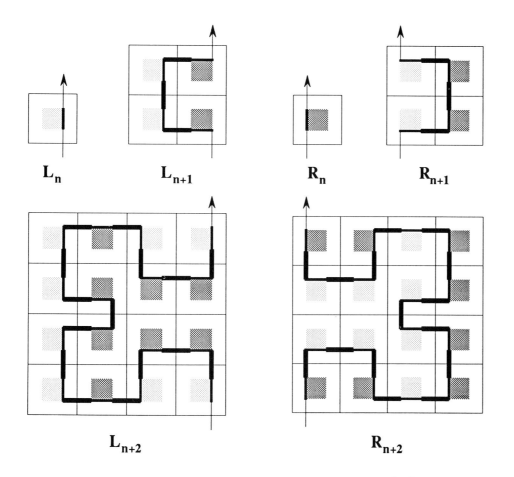

Figure 1.15: Recursive construction of the Hilbert curve [63] in terms of node replacement

and rotated in order to align its entry point P_A and direction $\vec{p}_A$ with the current position and orientation of the turtle. Having placed A, the turtle is assigned the resulting exit position Q_A and direction $\vec{q}_A$.

For example, assuming that the contact points and directions of subfigures L_n and R_n are as in Figure 1.15, the figures L_{n+1} and R_{n+1} are captured by the following formulas:

Recursive formulas

$$
\begin{aligned}
L_{n+1} &= +R_nF - L_nFL_n - FR_n+ \\
R_{n+1} &= -L_nF + R_nFR_n + FL_n-
\end{aligned}
$$

Suppose that curves L_0 and R_0 are given. One way of evaluating the string L_n (or R_n) for $n > 0$ is to generate successive strings recursively, in the order of decreasing value of index n. For example, the computation of L_2 would proceed as follows:

$$
\begin{aligned}
L_2 &= +R_1F - L_1FL_1 - FR_1+ \\
&= +(-L_0F + R_0FR_0 + FL_0-)F - (+R_0F - L_0FL_0 - FR_0+) \\
&\quad F(+R_0F - L_0FL_0 - FR_0+) - F(-L_0F + R_0FR_0 + FL_0-)+
\end{aligned}
$$

1.4.3   Relationship between edge and node rewriting

The classes of curves that can be generated using the edge-rewriting and node-rewriting techniques are not disjoint. For example, reconsider the L-system that generates the dragon curve using edge replacement:

$$
\begin{aligned}
\omega &: \ F_l \\
p_1 &: \ F_l \to F_l + F_r+ \\
p_2 &: \ F_r \to -F_l - F_r
\end{aligned}
$$

Assume temporarily that a production predecessor can contain more than one letter; thus an entire subword can be replaced by the successor of a single production (a formalization of this concept is termed *pseudo-L-systems* [109]). The dragon-generating L-system can be rewritten as:

$$
\begin{aligned}
\omega &: \ Fl \\
p_1 &: \ Fl \to Fl + rF+ \\
p_2 &: \ rF \to -Fl - rF
\end{aligned}
$$

where the symbols l and r are not interpreted by the turtle. Production p_1 replaces the letter l by the string $l + rF-$ while the leading letter F is left intact. In a similar way, production p_2 replaces the letter r by the string $-Fl - r$ and leaves the trailing F intact. Thus, the L-system can be transformed into node-rewriting form as follows:

$$
\begin{aligned}
\omega &: \ Fl \\
p_1 &: \ l \to l + rF+ \\
p_2 &: \ r \to -Fl - r
\end{aligned}
$$

In practice, the choice between edge rewriting and node rewriting is often a matter of convenience. Neither approach offers an automatic, general method for constructing L-systems that capture given structures. However, the distinction between edge and node rewriting makes it easier to understand the intricacies of L-system operation, and in this sense assists in the modeling task. Specifically, the problem of filling a region by a self-avoiding curve is biologically relevant, since some plant structures, such as leaves, may tend to fill a plane without overlapping [38, 66, 67, 94].

1.5   Modeling in three dimensions

Turtle interpretation of L-systems can be extended to three dimensions following the ideas of Abelson and diSessa [1]. The key concept is to represent the current *orientation* of the turtle in space by three vectors $\vec{H}, \vec{L}, \vec{U}$, indicating the turtle's *heading*, the direction to the *left*, and the direction *up*. These vectors have unit length, are perpendicular to each

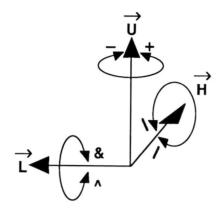

Figure 1.18: Controlling the turtle in three dimensions

other, and satisfy the equation $\vec{H} \times \vec{L} = \vec{U}$. Rotations of the turtle are then expressed by the equation

$$\begin{bmatrix} \vec{H}' & \vec{L}' & \vec{U}' \end{bmatrix} = \begin{bmatrix} \vec{H} & \vec{L} & \vec{U} \end{bmatrix} \mathbf{R},$$

where $\mathbf{R}$ is a 3×3 rotation matrix [40]. Specifically, rotations by angle α about vectors $\vec{U}, \vec{L}$ and $\vec{H}$ are represented by the matrices:

$$\mathbf{R_U}(\alpha) = \begin{bmatrix} \cos\alpha & \sin\alpha & 0 \\ -\sin\alpha & \cos\alpha & 0 \\ 0 & 0 & 1 \end{bmatrix}$$

$$\mathbf{R_L}(\alpha) = \begin{bmatrix} \cos\alpha & 0 & -\sin\alpha \\ 0 & 1 & 0 \\ \sin\alpha & 0 & \cos\alpha \end{bmatrix}$$

$$\mathbf{R_H}(\alpha) = \begin{bmatrix} 1 & 0 & 0 \\ 0 & \cos\alpha & -\sin\alpha \\ 0 & \sin\alpha & \cos\alpha \end{bmatrix}$$

The following symbols control turtle orientation in space (Figure 1.18):

+   Turn left by angle δ, using rotation matrix $\mathbf{R_U}(\delta)$.

−   Turn right by angle δ, using rotation matrix $\mathbf{R_U}(-\delta)$.

&   Pitch down by angle δ, using rotation matrix $\mathbf{R_L}(\delta)$.

∧   Pitch up by angle δ, using rotation matrix $\mathbf{R_L}(-\delta)$.

\   Roll left by angle δ, using rotation matrix $\mathbf{R_H}(\delta)$.

/   Roll right by angle δ, using rotation matrix $\mathbf{R_H}(-\delta)$.

| Turn around, using rotation matrix $\mathbf{R_U}(180°)$.

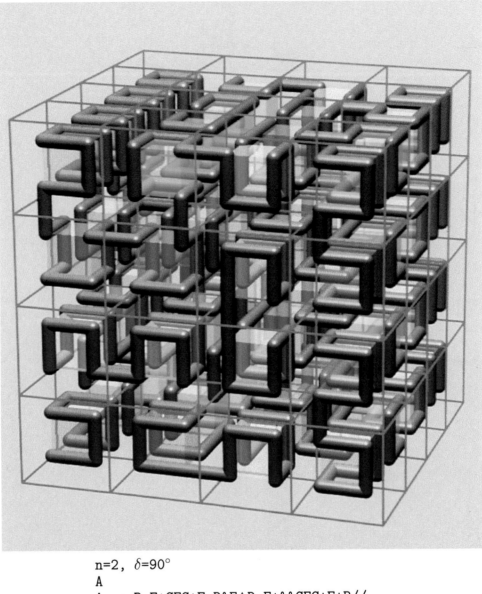

n=2, δ=90°

A

A → B-F+CFC+F-D&F∧D-F+&&CFC+F+B//

B → A&F∧CFB∧F∧D∧∧-F-D∧|F∧B|FC∧F∧A//

C → |D∧|F∧B-F+C∧F∧A&&FA&F∧C+F+B∧F∧D//

D → |CFB-F+B|FA&F∧A&&FB-F+B|FC//

Figure 1.19: A three-dimensional extension of the Hilbert curve [139]. Colors represent three-dimensional "frames" associated with symbols A (red), B (blue), C (green) and D (yellow).

As an example of a three-dimensional object created using an L-system, consider the extension of the Hilbert curve shown in Figure 1.19. The L-system was constructed with the node-replacement technique discussed in the previous section, using cubes and "macrocubes" instead of tiles and macrotiles.

1.6   Branching structures

According to the rules presented so far, the turtle interprets a character string as a sequence of line segments. Depending on the segment lengths and the angles between them, the resulting line is self-intersecting or not, can be more or less convoluted, and may have some segments drawn many times and others made invisible, but it always remains just a single line. However, the plant kingdom is dominated by branching structures; thus a mathematical description of tree-like shapes and methods for generating them are needed for modeling purposes. An axial tree [89, 117] complements the graph-theoretic notion of a rooted tree [108] with the botanically motivated notion of branch axis.

1.6.1   Axial trees

A *rooted tree* has edges that are labeled and directed. The edge sequences form paths from a distinguished node, called the *root* or *base*, to the *terminal nodes*. In the biological context, these edges are referred to as *branch segments*. A segment followed by at least one more segment in some path is called an *internode*. A terminal segment (with no succeeding edges) is called an *apex*.

An *axial tree* is a special type of rooted tree (Figure 1.20). At each of its nodes, at most one outgoing *straight* segment is distinguished. All remaining edges are called *lateral* or *side* segments. A sequence of segments is called an *axis* if:

- the first segment in the sequence originates at the root of the tree or as a lateral segment at some node,

- each subsequent segment is a straight segment, and

- the last segment is not followed by any straight segment in the tree.

Together with all its descendants, an axis constitutes a *branch*. A branch is itself an axial (sub)tree.

Axes and branches are ordered. The axis originating at the root of the entire plant has order zero. An axis originating as a lateral segment of an n-order parent axis has order $n+1$. The order of a branch is equal to the order of its lowest-order or *main* axis.

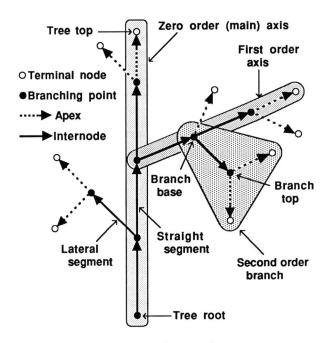

Figure 1.20: An axial tree

Figure 1.21: Sample tree generated using a method based on Horton–Strahler analysis of branching patterns

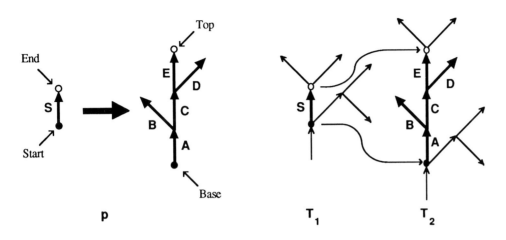

Figure 1.22: A tree production p and its application to the edge S in a tree T_1

Axial trees are purely topological objects. The geometric connotation of such terms as straight segment, lateral segment and axis should be viewed at this point as an intuitive link between the graph-theoretic formalism and real plant structures.

The proposed scheme for ordering branches in axial trees was introduced originally by Gravelius [53]. MacDonald [94, pages 110–121] surveys this and other methods applicable to biological and geographical data such as stream networks. Of special interest are methods proposed by Horton [70, 71] and Strahler, which served as a basis for synthesizing botanical trees [37, 152] (Figure 1.21).

1.6.2 Tree OL-systems

In order to model development of branching structures, a rewriting mechanism can be used that operates directly on axial trees. A rewriting rule, or *tree production*, replaces a predecessor edge by a successor axial tree in such a way that the starting node of the predecessor is identified with the successor's base and the ending node is identified with the successor's top (Figure 1.22).

A *tree OL-system G* is specified by three components: a set of edge labels V, an initial tree ω with labels from V, and a set of tree productions P. Given the L-system G, an axial tree T_2 is directly derived from a tree T_1, noted $T_1 \Rightarrow T_2$, if T_2 is obtained from T_1 by simultaneously replacing each edge in T_1 by its successor according to the production set P. A tree T is generated by G in a derivation of length n if there exists a sequence of trees $T_0, T_1, \ldots, T_n$ such that $T_0 = \omega, T_n = T$ and $T_0 \Rightarrow T_1 \Rightarrow \ldots \Rightarrow T_n$.

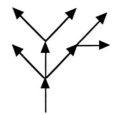

F[+F][-F[-F]F]F[+F][-F]

Figure 1.23: Bracketed string representation of an axial tree

1.6.3   Bracketed OL-systems

The definition of tree L-systems does not specify the data structure for representing axial trees. One possibility is to use a list representation with a tree topology. Alternatively, axial trees can be represented using *strings with brackets* [82]. A similar distinction can be observed in Koch constructions, which can be implemented either by rewriting edges and polygons or their string representations. An extension of turtle interpretation to strings with brackets and the operation of bracketed L-systems [109, 111] are described below.

Two new symbols are introduced to delimit a branch. They are interpreted by the turtle as follows:

Stack operations

[Push the current state of the turtle onto a pushdown stack. The information saved on the stack contains the turtle's position and orientation, and possibly other attributes such as the color and width of lines being drawn.

]   Pop a state from the stack and make it the current state of the turtle. No line is drawn, although in general the position of the turtle changes.

An example of an axial tree and its string representation are shown in Figure 1.23.

2D structures

Derivations in bracketed OL-systems proceed as in OL-systems without brackets. The brackets replace themselves. Examples of two-dimensional branching structures generated by bracketed OL-systems are shown in Figure 1.24.

Bush-like structure

Figure 1.25 is an example of a three-dimensional bush-like structure generated by a bracketed L-system. Production p_1 creates three new branches from an apex of the old branch. A branch consists of an edge F forming the initial internode, a leaf L and an apex A (which will subsequently create three new branches). Productions p_2 and p_3

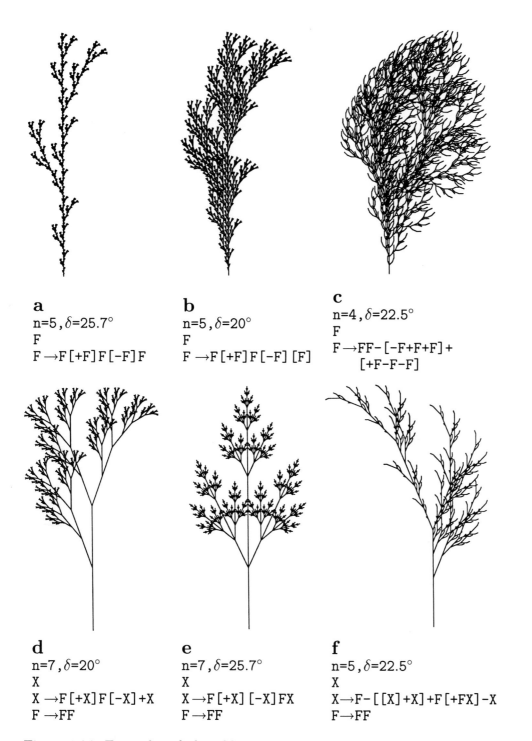

a
n=5,δ=25.7°
F
F →F[+F]F[-F]F

b
n=5,δ=20°
F
F →F[+F]F[-F][F]

c
n=4,δ=22.5°
F
F →FF-[-F+F+F]+
      [+F-F-F]

d
n=7,δ=20°
X
X →F[+X]F[-X]+X
F →FF

e
n=7,δ=25.7°
X
X →F[+X][-X]FX
F →FF

f
n=5,δ=22.5°
X
X→F-[[X]+X]+F[+FX]-X
F→FF

Figure 1.24: Examples of plant-like structures generated by bracketed OL-systems. L-systems (a), (b) and (c) are edge-rewriting, while (d), (e) and (f) are node-rewriting.

n=7, δ=22.5°

```
ω  :  A
p₁ :  A → [&FL!A]/////'[&FL!A]///////'[&FL!A]
p₂ :  F → S ///// F
p₃ :  S → F L
p₄ :  L → [''' ∧∧{-f+f+f-|-f+f+f}]
```

Figure 1.25: A three-dimensional bush-like structure

specify internode growth. In subsequent derivation steps the internode gets longer and acquires new leaves. This violates a biological rule of *subapical growth* (discussed in detail in Chapter 3), but produces an acceptable visual effect in a still picture. Production p_4 specifies the leaf as a filled polygon with six edges. Its boundary is formed from the edges f enclosed between the braces { and } (see Chapter 5 for further discussion). The symbols ! and ′ are used to decrement the diameter of segments and increment the current index to the color table, respectively.

Plant with flowers

Another example of a three-dimensional plant is shown in Figure 1.26. The L-system can be described and analyzed in a way similar to the previous one.

n=5, δ=18°

ω : **plant**
p_1 : **plant** $\to$ **internode** + [**plant** + **flower**] $- - //$
        [$- -$ **leaf**] **internode** [+ + **leaf**] $-$
        [**plant flower**] + + **plant flower**
p_2 : **internode** $\to$ F **seg** [// & & **leaf**] [// $\wedge$ $\wedge$ **leaf**] F **seg**
p_3 : **seg** $\to$ **seg** F **seg**
p_4 : **leaf** $\to$ [' { +f–ff–f+ | +f–ff–f }]
p_5 : **flower** $\to$ [& & & **pedicel** ' / **wedge** //// **wedge** ////
        wedge //// **wedge** //// **wedge**]
p_6 : **pedicel** $\to$ FF
p_7 : **wedge** $\to$ [' $\wedge$ F] [{ & & & & –f+f | –f+f }]

Figure 1.26: A plant generated by an L-system

1.7   Stochastic L-systems

All plants generated by the same deterministic L-system are identical. An attempt to combine them in the same picture would produce a striking, artificial regularity. In order to prevent this effect, it is necessary to introduce specimen-to-specimen variations that will preserve the general aspects of a plant but will modify its details.

Variation can be achieved by randomizing the turtle interpretation, the L-system, or both. Randomization of the interpretation alone has a limited effect. While the geometric aspects of a plant — such as the stem lengths and branching angles — are modified, the underlying topology remains unchanged. In contrast, stochastic application of productions may affect both the topology and the geometry of the plant. The following definition is similar to that of Yokomori [162] and Eichhorst and Savitch [35].

L-system   A *stochastic 0L-system* is an ordered quadruplet $G_\pi = \langle V, \omega, P, \pi \rangle$. The alphabet V, the axiom ω and the set of productions P are defined as in an 0L-system (page 4). Function $\pi : P \to (0, 1]$, called the *probability distribution*, maps the set of productions into the set of *production probabilities*. It is assumed that for any letter $a \in V$, the sum of probabilities of all productions with the predecessor a is equal to 1.

Derivation   We will call the derivation $\mu \Rightarrow \nu$ a *stochastic derivation* in G_π if for each *occurrence* of the letter a in the word μ the probability of applying production p with the predecessor a is equal to $\pi(p)$. Thus, different productions with the same predecessor can be applied to various occurrences of the same letter in one derivation step.

Example   A simple example of a stochastic L-system is given below.

$$
\begin{aligned}
\omega \;&: \; F \\
p_1 \;&: \; F \overset{.33}{\to} F[+F]F[-F]F \\
p_2 \;&: \; F \overset{.33}{\to} F[+F]F \\
p_3 \;&: \; F \overset{.34}{\to} F[-F]F
\end{aligned}
$$

The production probabilities are listed above the derivation symbol $\to$. Each production can be selected with approximately the same probability of 1/3. Examples of branching structures generated by this L-system with derivations of length 5 are shown in Figure 1.27. Note that these structures look like different specimens of the same (albeit fictitious) plant species.

Flower field   A more complex example is shown in Figure 1.28. The field consists of four rows and four columns of plants. All plants are generated by a stochastic modification of the L-system used to generate Figure 1.26.

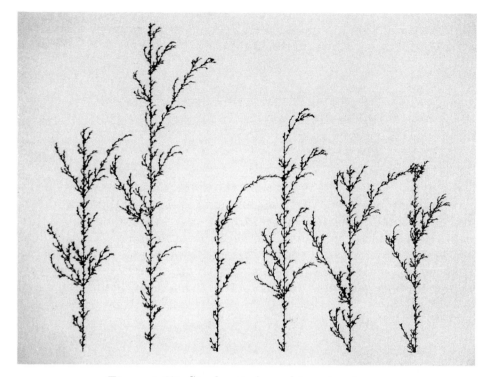

Figure 1.27: Stochastic branching structures

Figure 1.28: Flower field

The essence of this modification is to replace the original production p_3 by the following three productions:

$$p_3' : \quad \textbf{seg} \overset{.33}{\to} \textbf{seg} \, [\, // \, \& \, \& \, \textbf{leaf} \,] \, [// \, \wedge\wedge \, \textbf{leaf} \,] \, F \, \textbf{seg}$$
$$p_3'' : \quad \textbf{seg} \overset{.33}{\to} \textbf{seg} \, F \, \textbf{seg}$$
$$p_3''' : \quad \textbf{seg} \overset{.34}{\to} \textbf{seg}$$

Thus, in any step of the derivation, the stem segment **seg** may either grow and produce new leaves (production p_3'), grow without producing new leaves (production p_3''), or not grow at all (production p_3'''). All three events occur with approximately the same probability. The resulting field appears to consist of various specimens of the same plant species. If the same L-system was used again (with different seed values for the random number generator), a variation of this image would be obtained.

1.8   Context-sensitive L-systems

Context in string L-systems

Productions in OL-systems are context-free; i.e. applicable regardless of the context in which the predecessor appears. However, production application may also depend on the predecessor's context. This effect is useful in simulating interactions between plant parts, due for example to the flow of nutrients or hormones. Various context-sensitive extensions of L-systems have been proposed and studied thoroughly in the past [62, 90, 128]. *2L-systems* use productions of the form $a_l < a > a_r \to \chi$, where the letter a (called the *strict predecessor*) can produce word χ if and only if a is preceded by letter a_l and followed by a_r. Thus, letters a_l and a_r form the left and the right *context* of a in this production. Productions in *1L-systems* have one-sided context only; consequently, they are either of the form $a_l < a \to \chi$ or $a > a_r \to \chi$. OL-systems, 1L-systems and 2L-systems belong to a wider class of *IL-systems*, also called *(k,l)-systems*. In a (k,l)-system, the left context is a word of length k and the right context is a word of length l.

In order to keep specifications of L-systems short, the usual notion of IL-systems has been modified here by allowing productions with different context lengths to coexist within a single system. Furthermore, context-sensitive productions are assumed to have precedence over context-free productions with the same strict predecessor. Consequently, if a context-free and a context-sensitive production both apply to a given letter, the context-sensitive one should be selected. If no production applies, this letter is replaced by itself as previously assumed for OL-systems.

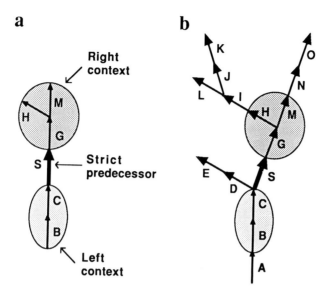

Figure 1.29: The predecessor of a context-sensitive tree production (a) matches edge S in a tree T (b)

The following sample 1L-system makes use of context to simulate signal propagation throughout a string of symbols:

Signal propagation

$$\omega \; : \quad baaaaaaaa$$
$$p_1 : \quad b < a \;\; \rightarrow b$$
$$p_2 : \qquad\quad b \;\; \rightarrow a$$

The first few words generated by this L-system are given below:

$$baaaaaaaa$$
$$abaaaaaaa$$
$$aabaaaaaa$$
$$aaabaaaaa$$
$$aaaabaaaa$$
$$\cdots$$

The letter b moves from the left side to the right side of the string.

A context-sensitive extension of tree L-systems requires neighbor edges of the replaced edge to be tested for context matching. A predecessor of a context-sensitive production p consists of three components: a *path l* forming the left context, an *edge S* called the strict predecessor, and an *axial tree r* constituting the right context (Figure 1.29). The asymmetry between the left context and the right context reflects the fact that there is only one path from the root of a tree to a given edge, while there can be many paths from this edge to various terminal nodes. Production p matches a given occurrence of the edge S in a tree T if l is a path in T terminating at the starting node of S, and r is a subtree

Context in tree L-systems

1.9   Growth functions

Exponential growth

During the synthesis of a plant model it is often convenient to distinguish productions that specify the branching pattern from those that describe elongation of plant segments. This separation can be observed in some of the L-systems considered so far. For example, in L-systems (d), (e) and (f) from Figure 1.24 the first productions capture the branching patterns, while the remaining productions, equal in all cases to $F \rightarrow FF$, describe elongation of segments represented by sequences of symbols F. The number of letters F in a string χ_n originating from a single letter F is doubled in each derivation step, thus the elongation is exponential, with $\text{length}(\chi_n) = 2^n$.

Basic properties

A function that describes the number of symbols in a word in terms of its derivation length is called a *growth function*. The theory of L-systems contains an extensive body of results on growth functions [62, 127]. The central observation is that the growth functions of DOL-systems are independent of the letter ordering in the productions and derived words. Consequently, the relation between the number of letter occurrences in a pair of words μ and ν, such that $\mu \Rightarrow \nu$, can be conveniently expressed using matrix notation.

Let $G = \langle V, \omega, P \rangle$ be a DOL-system and assume that letters of the alphabet V have been ordered, $V = \{a_1, a_2, \ldots, a_m\}$. Construct a square matrix $Q_{m \times m}$, where entry q_{ij} is equal to the number of occurrences of letter a_j in the successor of the production with predecessor a_i. Let $\underline{a}_i^k$ denote the number of occurrences of letter a_i in the word x generated by G in a derivation of length k. The definition of direct derivation in a DOL-system implies that

$$\begin{bmatrix} \underline{a}_1^k & \underline{a}_2^k & \cdots & \underline{a}_m^k \end{bmatrix} \begin{bmatrix} q_{11} & q_{12} & \cdots & q_{1m} \\ q_{21} & q_{22} & \cdots & q_{2m} \\ \vdots & & & \\ q_{m1} & q_{m2} & \cdots & q_{mm} \end{bmatrix} = \begin{bmatrix} \underline{a}_1^{k+1} & \underline{a}_2^{k+1} & \cdots & \underline{a}_m^{k+1} \end{bmatrix}.$$

This matrix notation is useful in the analysis of growth functions. For example, consider the following L-system:

$$\begin{array}{rl} \omega : & a \\ p_1 : & a \rightarrow ab \\ p_2 : & b \rightarrow a \end{array} \qquad (1.2)$$

The relationship between the number of occurrences of letters a and b in two consecutively derived words is

$$\begin{bmatrix} \underline{a}^k & \underline{b}^k \end{bmatrix} \begin{bmatrix} 1 & 1 \\ 1 & 0 \end{bmatrix} = \begin{bmatrix} \underline{a}^{k+1} & \underline{b}^{k+1} \end{bmatrix}$$

or

$$\underline{a}^{k+1} = \underline{a}^k + \underline{b}^k = \underline{a}^k + \underline{a}^{k-1}$$

for $k = 1, 2, 3, \ldots$. From the axiom it follows that $\underline{a}^0 = 1$ and $\underline{a}^1 = \underline{b}^0 = 0$. Thus, the number of letters a in the strings generated by the L-system specified in equation (1.2) grows according to the Fibonacci series: $1, 1, 2, 3, 5, 8, \ldots$. This growth function was implemented by productions p_2 and p_3 in the L-system generating the bush in Figure 1.25 (page 26) to describe the elongation of its internodes.

Polynomial growth

Polynomial growth functions of arbitrary degree can be obtained using L-systems of the following form:

$$
\begin{aligned}
\omega &: \quad a_0 \\
p_1 &: \quad a_0 \rightarrow a_0 a_1 \\
p_2 &: \quad a_1 \rightarrow a_1 a_2 \\
p_3 &: \quad a_2 \rightarrow a_2 a_3 \\
p_4 &: \quad a_3 \rightarrow a_3 a_4 \\
&\quad \vdots
\end{aligned}
$$

The matrix Q is given below:

$$
Q = \begin{bmatrix}
1 & 1 & 0 & 0 & \cdots \\
0 & 1 & 1 & 0 & \cdots \\
0 & 0 & 1 & 1 & \cdots \\
0 & 0 & 0 & 1 & \cdots \\
& & \vdots &
\end{bmatrix}
$$

Thus, for any $i, k \geq 1$, the number $\underline{a}_i^k$ of occurrences of symbol a_i in the string generated in a derivation of length k satisfies the equality

$$\underline{a}_i^k + \underline{a}_{i+1}^k = \underline{a}_{i+1}^{k+1}.$$

Taking into consideration the axiom, the distribution of letters a_i as a function of the derivation length is captured by the following table (only non-zero terms are shown):

k	$\underline{a}_0^k$	$\underline{a}_1^k$	$\underline{a}_2^k$	$\underline{a}_3^k$	$\underline{a}_4^k$	$\underline{a}_5^k$	$\underline{a}_6^k$	$\underline{a}_7^k$
0	1							
1	1	1						
2	1	2	1					
3	1	3	3	1				
4	1	4	6	4	1			
5	1	5	10	10	5	1		
6	1	6	15	20	15	6	1	
7	1	7	21	35	35	21	7	1

$$\vdots$$

This table represents the Pascal triangle, thus for any $k \geq i \geq 1$ its terms satisfy the following equality:

$$\underline{a}_i^k = \left(\begin{array}{c} k \\ i \end{array} \right) = \frac{k(k-1)\cdots(k-i+1)}{1 \cdot 2 \cdots i}$$

Consequently, the number of occurrences of letter a_i as a function of the derivation length k is expressed by a polynomial of degree i. By identifying letter a_i with the turtle symbol F, it is possible to model internode elongation expressed by polynomials of arbitrary degree $i \geq 0$. This observation was generalized by Szilard [140], who developed an algorithm for constructing a DOL-system with growth functions specified by any positive, nondecreasing polynomials with integer coefficients [62, page 276].

Characterization   The examples of growth functions considered so far include exponential and polynomial functions. Rozenberg and Salomaa [127, pages 30–38] show that, in general, the growth function $f_G(n)$ of any DOL-system $G = \langle V, \omega, P \rangle$ is a combination of polynomial and exponential functions:

$$f_G(n) = \sum_{i=1}^{s} P_i(n)\rho_i^n \quad \text{for} \quad n \geq n_0, \tag{1.3}$$

where $P_i(n)$ denotes a polynomial with integer coefficients, ρ_i is a non-negative integer, and n_0 is the total number of letters in the alphabet V. Unfortunately, many growth processes observed in nature cannot be described by equation (1.3). Two approaches are then possible within the framework of the theory of L-systems.

Sigmoidal growth   The first is to extend the size n_0 of the alphabet V, so that the growth process of interest will be captured by the initial derivation steps, $\omega = \mu_0 \Rightarrow \mu_1 \Rightarrow \cdots \Rightarrow \mu_{n_0}$, before equation (1.3) starts to apply. For example, the L-system

$$\begin{array}{rll}
\omega & : & a_0 \\
p_i & : & a_i \rightarrow a_{i+1}b_0 \quad \text{for} \quad i < k \\
p_{k+j} & : & b_j \rightarrow b_{j+1}F \quad \text{for} \quad j < l
\end{array} \tag{1.4}$$

over the alphabet $V = \{a_0, a_1, ..., a_k\} \cup \{b_0, b_1, ..., b_l\} \cup \{F\}$ can be used to approximate a sigmoidal elongation of a segment represented by a sequence of symbols F (Figure 1.32). The term *sigmoidal* refers to a function with a plot in the shape of the letter S. Such functions are commonly found in biological processes [143], with the initial part of the curve representing the growth of a young organism, and the latter part corresponding to the organism close to its final size.

Square-root growth   The second approach to the synthesis of growth functions outside the class captured by equation (1.3) is to use context-sensitive L-systems. For example, the following 2L-system has a growth function given by $f_G(n) = \lfloor \sqrt{n} \rfloor + 4$, where $\lfloor x \rfloor$ is the floor function.

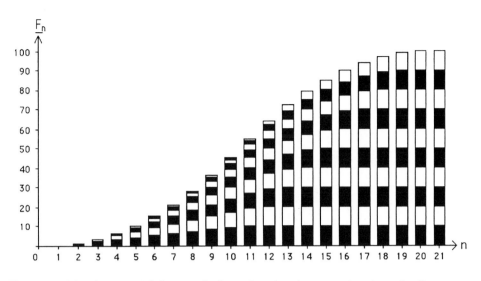

Figure 1.32: A sigmoidal growth function implemented using the L-system in equation (1.4), for $k = l = 20$

$$
\begin{array}{rlcl}
\omega : & XF_uF_aX \\
p_1 : & F_u < F_a > F_a & \rightarrow & F_u \\
p_2 : & F_u < F_a > X & \rightarrow & F_dF_a \\
p_3 : & F_a < F_a > F_d & \rightarrow & F_d \\
p_4 : & X < F_a > F_d & \rightarrow & F_u \\
p_5 : & F_u & \rightarrow & F_a \\
p_6 : & F_d & \rightarrow & F_a
\end{array}
\tag{1.5}
$$

The operation of this L-system is illustrated in Figure 1.33. Productions p_1 and p_3, together with p_5 and p_6, propagate symbols F_u and F_d up and down the string of symbols μ. Productions p_2 and p_4 change the propagation direction, after symbol X marking a string end has been reached by F_u or F_d, respectively. In addition, p_2 extends the string with a symbol F_a. Thus, the number of derivation steps increases by two between consecutive applications of production p_2. As a result, string extension occurs at derivation steps n expressed by the square of the string length, which yields the growth function $\lfloor \sqrt{n} \rfloor + 4$.

Limitations

In practice it is often difficult, if not impossible, to find L-systems with the required growth functions. Vitányi [153] illustrates this by referring to sigmoidal curves:

> If we want to obtain sigmoidal growth curves with the original L-systems then not even the introduction of cell interaction can help us out. In the first place, we end up constructing quite unlikely flows of messages through the organism, which are more suitable to electronic computers, and in fact give the organism universal computing power. Secondly, and this is more fundamental, we can not obtain

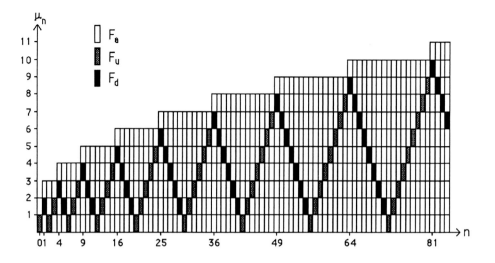

Figure 1.33: Square-root growth implemented using the L-system specified in equation (1.5)

growth which, always increasing the size of the organism, tends towards stability in the limit. The slowest increasing growth we can obtain by allowing cell interaction is logarithmic and thus can not at all account for the asymptotic behavior of sigmoidal growth functions.

In the next section we present an extension of L-systems that makes it possible to avoid this problem by allowing for explicit inclusion of growth functions into L-system specifications.

1.10   Parametric L-systems

Motivation

Although L-systems with turtle interpretation make it possible to generate a variety of interesting objects, from abstract fractals to plant-like branching structures, their modeling power is quite limited. A major problem can be traced to the reduction of all lines to integer multiples of the unit segment. As a result, even such a simple figure as an isosceles right-angled triangle cannot be traced exactly, since the ratio of its hypotenuse length to the length of a side is expressed by the irrational number $\sqrt{2}$. Rational approximation of line length provides only a limited solution, because the unit step must be the smallest common denominator of all line lengths in the modeled structure. Consequently, the representation of a simple plant module, such as an internode, may require a large number of symbols. The same argument applies to angles. Problems become even more pronounced while simulating changes to the modeled structure over time, since some growth functions cannot be expressed conveniently using L-systems. Generally, it is difficult to capture continuous phenomena, since the obvious technique of discretizing continuous values may require a large number of quantization levels, yielding L-systems with hundreds of symbols and productions. Consequently, model specification becomes difficult, and the mathematical beauty of L-systems is lost.

In order to solve similar problems, Lindenmayer proposed that numerical parameters be associated with L-system symbols [83]. He illustrated this idea by referring to the continuous development of branching structures and diffusion of chemical compounds in a nonbranching filament of *Anabaena catenula*. Both problems were revisited in later papers [25, 77]. A definition of parametric L-systems was formulated by Prusinkiewicz and Hanan [113] and is presented below.

1.10.1   Parametric OL-systems

Parametric L-systems operate on *parametric words*, which are strings of *modules* consisting of *letters* with associated *parameters*. The letters belong to an *alphabet* V, and the parameters belong to the set of *real numbers* $\Re$. A module with letter $A \in V$ and parameters $a_1, a_2, ..., a_n \in \Re$ is denoted by $A(a_1, a_2, ..., a_n)$. Every module belongs to the set $M = V \times \Re^*$, where $\Re^*$ is the set of all finite sequences of parameters. The set of all strings of modules and the set of all nonempty strings are denoted by $M^* = (V \times \Re^*)^*$ and $M^+ = (V \times \Re^*)^+$, respectively.

Parametric words

The real-valued *actual* parameters appearing in the words correspond with *formal* parameters used in the specification of L-system productions. If Σ is a set of formal parameters, then $C(\Sigma)$ denotes a *logical expression* with parameters from Σ, and $E(\Sigma)$ is an *arithmetic expression* with parameters from the same set. Both types of expressions consist of formal parameters and numeric constants, combined using the arithmetic operators $+, -, *, /$; the exponentiation operator $\wedge$, the relational operators $<, >, =$; the logical operators $!, \&, |$ (not, and, or); and parentheses (). Standard rules for constructing syntactically correct expressions and for operator precedence are observed. Relational and logical expressions evaluate to zero for false and one for true. A logical statement specified as the empty string is assumed to have value one. The sets of all correctly constructed logical and arithmetic expressions with parameters from Σ are noted $\mathcal{C}(\Sigma)$ and $\mathcal{E}(\Sigma)$.

Expressions

A *parametric OL-system* is defined as an ordered quadruplet $G = \langle V, \Sigma, \omega, P \rangle$, where

Parametric OL-system

- V is the *alphabet* of the system,

- Σ is the *set of formal parameters*,

- $\omega \in (V \times \Re^*)^+$ is a nonempty parametric word called the *axiom*,

- $P \subset (V \times \Sigma^*) \times \mathcal{C}(\Sigma) \times (V \times \mathcal{E}(\Sigma))^*$ is a finite *set of productions*.

The symbols : and $\rightarrow$ are used to separate the three components of a production: the *predecessor*, the *condition* and the *successor*. For example, a production with predecessor $A(t)$, condition $t > 5$ and successor $B(t+1)CD(t \wedge 0.5, t-2)$ is written as

$$A(t) : t > 5 \rightarrow B(t+1)CD(t \wedge 0.5, t-2). \qquad (1.6)$$

Derivation
A production *matches* a module in a parametric word if the following conditions are met:

- the letter in the module and the letter in the production predecessor are the same,

- the number of actual parameters in the module is equal to the number of formal parameters in the production predecessor, and

- the condition evaluates to *true* if the actual parameter values are substituted for the formal parameters in the production.

A matching production can be *applied* to the module, creating a string of modules specified by the production successor. The actual parameter values are substituted for the formal parameters according to their position. For example, production (1.6) above matches a module $A(9)$, since the letter A in the module is the same as in the production predecessor, there is one actual parameter in the module $A(9)$ and one formal parameter in the predecessor $A(t)$, and the logical expression $t > 5$ is true for $t = 9$. The result of the application of this production is a parametric word $B(10)CD(3,7)$.

If a module a produces a parametric word χ as the result of a production application in an L-system G, we write $a \mapsto \chi$. Given a parametric word $\mu = a_1 a_2 ... a_m$, we say that the word $\nu = \chi_1 \chi_2 ... \chi_m$ is *directly derived* from (or *generated* by) μ and write $\mu \Longrightarrow \nu$ if and only if $a_i \mapsto \chi_i$ for all $i = 1, 2, ..., m$. A parametric word ν is generated by G in a *derivation of length* n if there exists a sequence of words $\mu_0, \mu_1, ..., \mu_n$ such that $\mu_0 = \omega$, $\mu_n = \nu$ and $\mu_0 \Longrightarrow \mu_1 \Longrightarrow ... \Longrightarrow \mu_n$.

Example
An example of a parametric L-system is given below.

$$
\begin{array}{llll}
\omega : & B(2)A(4,4) & & \\
p_1 : & A(x,y) & : y <= 3 & \rightarrow \quad A(x*2, x+y) \\
p_2 : & A(x,y) & : y > 3 & \rightarrow \quad B(x)A(x/y, 0) \\
p_3 : & B(x) & : x < 1 & \rightarrow \quad C \\
p_4 : & B(x) & : x >= 1 & \rightarrow \quad B(x-1)
\end{array} \qquad (1.7)
$$

As in the case of non-parametric L-systems, it is assumed that a module replaces itself if no matching production is found in the set P. The words obtained in the first few derivation steps are shown in Figure 1.34.

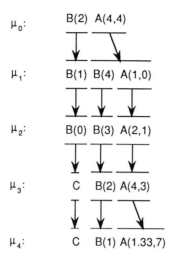

Figure 1.34: The initial sequence of strings generated by the parametric L-system specified in equation (1.7)

1.10.2 Parametric 2L-systems

Productions in parametric OL-systems are context-free, *i.e.*, applicable regardless of the context in which the predecessor appears. A context-sensitive extension is necessary to model information exchange between neighboring modules. In the parametric case, each component of the production predecessor (the left context, the strict predecessor and the right context) is a parametric word with letters from the alphabet V and formal parameters from the set Σ. Any formal parameters may appear in the condition and the production successor.

A sample context-sensitive production is given below: *Example*

$$A(x) < B(y) > C(z) : x + y + z > 10 \rightarrow E((x+y)/2)F((y+z)/2)$$

It can be applied to the module $B(5)$ that appears in a parametric word

$$\cdots A(4)B(5)C(6)\cdots \qquad (1.8)$$

since the sequence of letters A, B, C in the production and in parametric word (1.8) are the same, the numbers of formal parameters and actual parameters coincide, and the condition $4 + 5 + 6 > 10$ is true. As a result of the production application, the module $B(5)$ will be replaced by a pair of modules $E(4.5)F(5.5)$. Naturally, the modules $A(4)$ and $C(6)$ will be replaced by other productions in the same derivation step.

Parametric 2L-systems provide a convenient tool for expressing de-   *Anabaena with*
velopmental models that involve diffusion of substances throughout an   *heterocysts*
organism. A good example is provided by an extended model of the pattern of cells observed in *Anabaena catenula* and other blue-green bacteria [99]. This model was proposed by de Koster and Linden-mayer [25].

```
#define CH 900  /* high concentration */
#define CT 0.4  /* concentration threshold */
#define ST 3.9  /* segment size threshold */
#include H       /* heterocyst shape specification */
#ignore f ~ H
```

ω : -(90)F(0,0,CH)F(4,1,CH)F(0,0,CH)

p_1 : F(s,t,c) : t=1 & s>=6 →
                        F(s/3*2,2,c)f(1)F(s/3,1,c)

p_2 : F(s,t,c) : t=2 & s>=6 →
                        F(s/3,2,c)f(1)F(s/3*2,1,c)

p_3 : F(h,i,k) < F(s,t,c) > F(o,p,r) : s>ST|c>CT →
                        F(s+.1,t,c+0.25*(k+r-3*c))

p_4 : F(h,i,k) < F(s,t,c) > F(o,p,r) : !(s>ST|c>CT) →
                        F(0,0,CH) ~ H(1)

p_5 : H(s) : s<3 → H(s*1.1)

L-system 1.1: *Anabaena catenula*

Generally, the bacteria under consideration form a nonbranching filament consisting of two classes of cells: *vegetative cells* and *heterocysts*.
Usually, the vegetative cells divide and produce two daughter vegetative cells. This mechanism is captured by the L-system specified in
equation (1.1) and Figure 1.4 (page 5). However, in some cases the
vegetative cells differentiate into heterocysts. Their distribution forms
a well-defined pattern, characterized by a relatively constant number
of vegetative cells separating consecutive heterocysts. How does the
organism maintain constant spacing of heterocysts while growing? The
model explains this phenomenon using a biologically well-motivated
hypothesis that heterocyst distribution is regulated by nitrogen compounds produced by the heterocysts, transported from cell to cell across
the filament, and decayed in the vegetative cells. If the compound's
concentration in a young vegetative cell falls below a specific level, this
cell differentiates into a heterocyst (L-system 1.1).

The #define statements assign values to numerical constants used
in the L-system. The #include statement specifies the shape of a heterocyst (a disk) by referring to a library of predefined shapes (see Section 5.1). Cells are represented by modules $F(s,t,c)$, where s stands
for cell length, t is cell type (0 - heterocyst, 1 and 2 - vegetative types[1]),

---

[1]The model of *Anabaena* introduced in Section 1.2 distinguished between four
types of cells: a_r, b_r, a_l and b_l. Cells b do not divide and can be considered as young
forms of the corresponding cells a. Thus, the vegetative type 1 considered here
embraces cells a_r and b_r, while type 2 embraces cells a_l and b_l. The formal relationship
between the four-cell and two-cell models is further discussed in Chapter 6.

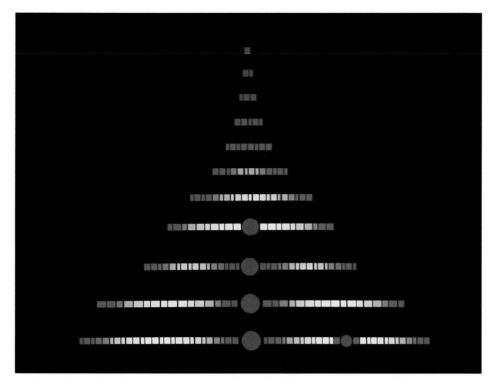

Figure 1.35: Development of *Anabaena catenula* with heterocysts, simulated using parametric L-system 1.1

and c represents the concentration of nitrogen compounds. Productions p_1 and p_2 describe division of the vegetative cells. They each create two daughter cells of unequal length. The difference between cells of type 1 and 2 lies in the ordering of the long and short daughter cells. Production p_3 captures the processes of transportation and decay of the nitrogen compounds. Their concentration c is related to the concentration in the neighboring cells k and r, and changes in each derivation step according to the formula

$$c' = c + 0.25(k + r - 3 * c).$$

Production p_4 describes differentiation of a vegetative cell into a heterocyst. The condition specifies that this process occurs when the cell length does not exceed the threshold value $ST = 3.9$ (which means that the cell is young enough), and the concentration of the nitrogen compounds falls below the threshold value $CT = 0.4$. Production p_5 describes the subsequent growth of the heterocyst.

Snapshots of the developmental sequence of *Anabaena* are given in Figure 1.35. The vegetative cells are shown as rectangles, colored according to the concentration of the nitrogen compounds (white means low concentration). The heterocysts are represented as red disks. The values of parameters CH, CT and ST were selected to provide the correct distribution of the heterocysts, and correspond closely to the values reported in [25]. The mathematical model made it possible to estimate these parameters, although they are not directly observable.

1.10.3   Turtle interpretation of parametric words

If one or more parameters are associated with a symbol interpreted by the turtle, the value of the first parameter controls the turtle's state. If the symbol is not followed by any parameter, default values specified outside the L-system are used as in the non-parametric case. The basic set of symbols affected by the introduction of parameters is listed below.

$F(a)$ Move forward a step of length $a > 0$. The position of the turtle changes to (x', y', z'), where $x' = x + a\vec{H}_x$, $y' = y + a\vec{H}_y$ and $z' = z + a\vec{H}_z$. A line segment is drawn between points (x, y, z) and (x', y', z').

$f(a)$ Move forward a step of length a without drawing a line.

$+(a)$ Rotate around $\vec{U}$ by an angle of a degrees. If a is positive, the turtle is turned to the left and if a is negative, the turn is to the right.

$\&(a)$ Rotate around $\vec{L}$ by an angle of a degrees. If a is positive, the turtle is pitched down and if a is negative, the turtle is pitched up.

$/(a)$ Rotate around $\vec{H}$ by an angle of a degrees. If a is positive, the turtle is rolled to the right and if a is negative, it is rolled to the left.

It should be noted that symbols $+$, $\&$, $\wedge$, and $/$ are used both as letters of the alphabet V and as operators in logical and arithmetic expressions. Their meaning depends on the context.

Row of trees

The following examples illustrate the operation of parametric L-systems with turtle interpretation. The first L-system is a coding of a Koch construction generating a variant of the snowflake curve (Figure 1.1 on page 2). The initiator (production predecessor) is the hypotenuse AB of a right triangle ABC (Figure 1.36). The first and the fourth edge of the generator subdivide AB into segments AD and DB, while the remaining two edges traverse the altitude CD in opposite directions. From elementary geometry it follows that the lengths of these segments satisfy the equations

$$q = c - p \qquad \text{and} \qquad h = \sqrt{pq}.$$

The edges of the generator can be associated with four triangles that are similar to ABC and tile it without gaps. According to the relationship between curve construction by edge rewriting and planar tilings

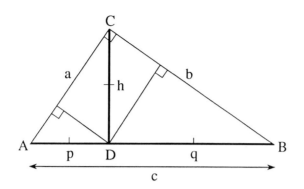

Figure 1.36: Construction of the generator for a "row of trees." The edges are associated with triangles indicated by ticks.

(Section 1.4.1), the generated curve will approximately fill the triangle ABC. The corresponding L-system is given below:

$$
\begin{aligned}
&\#\text{define}\quad c\quad 1 \\
&\#\text{define}\quad p\quad 0.3 \\
&\#\text{define}\quad q\quad c-p \\
&\#\text{define}\quad h\quad (p*q) \wedge 0.5
\end{aligned}
$$

$$
\begin{aligned}
\omega\ &:\ F(1) \\
p_1\ &:\ F(x) \rightarrow F(x*p) + F(x*h) - -F(x*h) + F(x*q)
\end{aligned}
$$

The resulting curve is shown in Figure 1.37a. In order to better visualize its structure, the angle increment has been set to 86° instead of 90°. The curve fills different areas with unequal density. This results from the fact that all edges, whether long or short, are replaced by the generator in every derivation step. A modified curve that fills the underlying triangle in a more uniform way is shown in Figure 1.37b. It was obtained by delaying the rewriting of shorter segments with respect to the longer ones, as specified by the following L-system.

$$
\begin{aligned}
\omega\ &:\ F(1,0) \\
p_1\ &:\ F(x,t):t=0\ \rightarrow\ F(x*p,2)+F(x*h,1)- \\
&\qquad\qquad\qquad\qquad\quad -F(x*h,1)+F(x*q,0) \\
p_2\ &:\ F(x,t):t>0\ \rightarrow\ F(x,t-1)
\end{aligned}
$$

The next example makes use of node rewriting (Section 1.4.2). The construction recursively subdivides a rectangular tile $ABCD$ into two tiles, $AEFD$ and $BCFE$, similar to $ABCD$ (Figure 1.38). The lengths of the edges form the proportion

Branching structure

$$
\frac{a}{b} = \frac{b}{\frac{1}{2}a},
$$

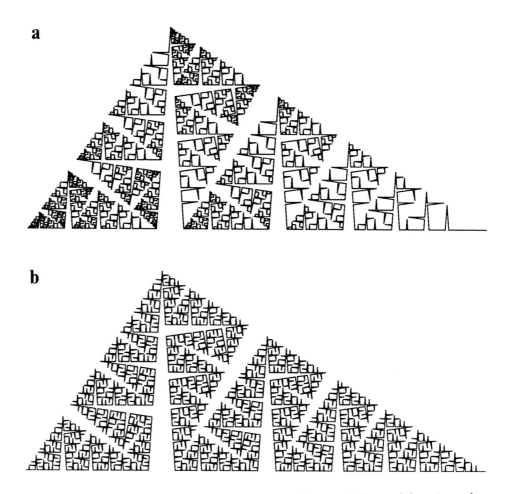

a

b

Figure 1.37: Two curves suggesting a "row of trees." Curve (b) is from [95, page 57].

which implies that $b = a/\sqrt{2}$. Each tile is associated with a single-point frame lying in the tile center. The tiles are connected by a branching line specified by the following L-system:

$$
\begin{aligned}
&\#\text{define } R \quad 1.456 \\
\omega :\ &A(1) \\
p_1 :\ &A(s) \rightarrow F(s)[+A(s/R)][-A(s/R)]
\end{aligned}
\qquad (1.9)
$$

The ratio of branch sizes R slightly exceeds the theoretical value of $\sqrt{2}$. As a result, the branching structure shown in Figure 1.39 is self-avoiding. The angle increment was set arbitrarily to $\delta = 85°$.

The L-system in equation (1.9) operates by appending segments of decreasing length to the structures obtained in previous derivation steps. Once a segment has been incorporated, its length does not change. A structure with identical proportions can be obtained by

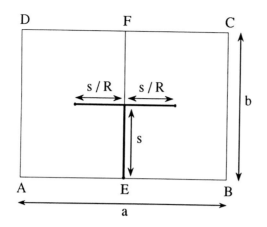

Figure 1.38: Tiling associated with a space-filling branching pattern

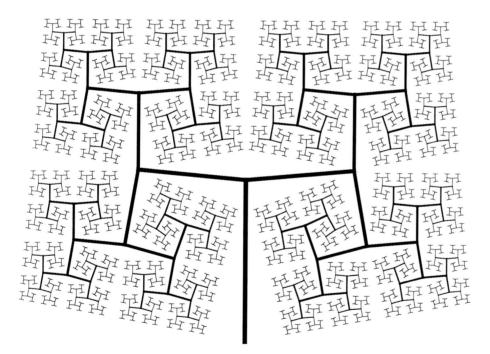

Figure 1.39: A branching pattern generated by the L-system specified in equation (1.9)

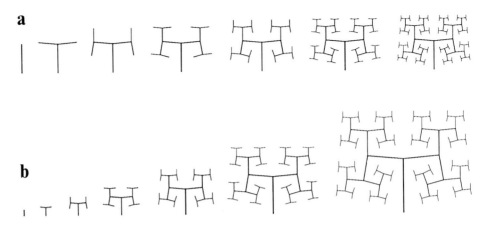

Figure 1.40: Initial sequences of figures generated by the L-systems specified in equations (1.9) and (1.10)

appending segments of constant length and increasing the lengths of previously created segments by constant R in each derivation step. The corresponding L-system is given below.

$$
\begin{aligned}
\omega &: \quad A \\
p_1 &: \quad A \quad\;\; \rightarrow F(1)[+A][-A] \\
p_2 &: \quad F(s) \;\; \rightarrow F(s*R)
\end{aligned}
\qquad (1.10)
$$

The initial sequence of structures obtained by both L-systems are compared in Figure 1.40. Sequence (a) emphasizes the fractal character of the resulting structure. Sequence (b) suggests the growth of a tree. The next two chapters show that this is not a mere coincidence, and the L-system specified in equation (1.10) is a simple, but in principle correct, developmental model of a *sympodial* branching pattern found in many herbaceous plants and trees.

Chapter 2

Modeling of trees

Computer simulation of branching patterns has a relatively long history. The first model was proposed by Ulam [149], (see also [138, pages 127–131]), and employed the concept of *cellular automata* that had been developed earlier by von Neumann and Ulam [156]. The branching pattern is created iteratively, starting with a single colored cell in a triangular grid, then coloring cells that touch one and only one vertex of a cell colored in the previous iteration step.

Cellular–space models

This basic idea gave rise to several extensions. Meinhardt [97, Chapter 15] substituted the triangular grid with a square one, and used the resulting cellular space to examine biological hypotheses related to the formation of net-like structures. In addition to pure branching patterns, his models capture the effect of branch reconnection or *anastomosis* that may take place between the veins of a leaf. Greene [54] extended cellular automata to three dimensions, and applied the resulting *voxel space automata* to simulate growth processes that react to the environment. For instance, Figure 2.1 presents the growth of a vine over a house. Cohen [15] simulated the development of a branching pattern using expansion rules that operate in a continuous "density field" rather than a discrete cellular or voxel space.

The common feature of these approaches is the emphasis on interactions between various elements of a growing structure, as well as the structure and the environment. Although interactions clearly influence the development of real plants, they also add to the complexity of the models. This may explain why simpler models, ignoring even such fundamental factors as collisions between branches, have been prevalent to date. The first model in that category was proposed by Honda [65] who studied the form of trees using the following assumptions (Figure 2.2).

Honda's model

- Tree segments are straight and their girth is not considered.

- A mother segment produces two daughter segments through one branching process.

Figure 2.1: *Organic architecture* by Greene [54]

- The lengths of the two daughter segments are shortened by constant ratios, r_1 and r_2, with respect to the mother segment.

- The mother segment and its two daughter segments are contained in the same *branch plane*. The daughter segments form constant *branching angles*, a_1 and a_2, with respect to the mother branch.

- The branch plane is fixed with respect to the direction of gravity so as to be closest to a horizontal plane.[1] An exception is made for branches attached to the main trunk. In this case, a constant *divergence angle* α between consecutively issued lateral segments is maintained (cf. Chapter 4).

By changing numerical parameters, Honda obtained a wide variety of tree-like shapes. With some improvements [38], his model was applied to investigate branching patterns of real trees [39, 66, 67, 68]. Subsequently, different rules for branching angles were proposed to capture the structure of trees in which planes of successive bifurcations are perpendicular to each other [69]. The results of Honda served as a basis for the tree models proposed by Aono and Kunii [2]. They suggested

---

[1]More formally, the line perpendicular to the mother segment and lying in the branch plane is horizontal.

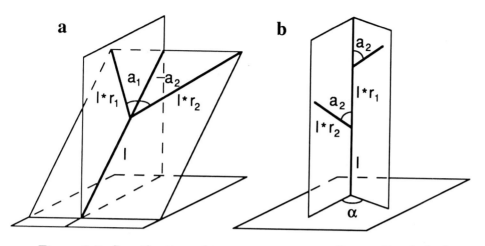

Figure 2.2: Specification of tree geometry according to Honda [65]

several extensions to his model, the most important of which was the biasing of branch positions in a particular direction, applied to produce the effects of wind, phototropism and gravity. A similar concept was introduced previously by Cohen [15], while more accurate physically-based methods for branch bending were developed by de Reffye [28] and Armstrong [4].

Realism

The models of Honda and Aono and Kunii were rendered using straight lines of constant or varying width to represent "tree skeletons." A substantial improvement in the realism of synthetic images was achieved by Bloomenthal [11] and Oppenheimer [105], who introduced curved branches, carefully modeled surfaces around branching points, and textured bark and leaves (Figure 2.3).

Stochastic models

The approaches stemming from the work of Honda defined branching structures using deterministic algorithms. In contrast, stochastic mechanisms are essential to the group of tree models proposed by Reeves and Blau [119], de Reffye et al. [30], and Remphrey, Neal and Steeves [120]. Although these models were described using different terminologies, they share the basic paradigm of specifying tree structures in terms of probabilities with which branches are formed. The work of Reeves and Blau aimed at producing tree-like shapes without delving into biological details of the modeled structures (Figure 2.4). In contrast, de Reffye et al. [29] used a stochastic approach to simulate the development of real plants by modeling the activity of buds at discrete time intervals. Given a clock signal, a bud can either:

Approach of de Reffye

- do nothing,

- become a flower,

- become an internode terminated by a new straight apex and one or more lateral apices subtended by leaves, or

- die and disappear.

Figure 2.3: *Acer graphics* by Bloomenthal [11]

Figure 2.4: A forest scene by Reeves [119] ©1984 Pixar

Figure 2.5: Oil palm tree canopy from CIRAD Modelisation Laboratory

These events occur according to stochastic laws characteristic for each variety and each species. The geometric parameters, such as the length and diameter of an internode, as well as branching angles, are also calculated according to stochastic laws.

The basic types of developmental rules incorporated into this method correspond to the 23 different types of tree architectures identified by Hallé, Oldeman and Tomlinson [58]. Detailed models of selected plant species were developed and are described in the literature [16, 20, 26, 27, 76]. A sample tree model is shown in Figure 2.5. The approach of Remphrey [120, 121, 122] is similar to that of de Reffye, except that larger time steps are used (one year in the model of bearberry described in [120]). Consequently, the stochastic rules must describe the entire configuration of lateral shoots that can be formed over a one-year period.

The application of L-systems to the generation of botanical trees was first considered by Aono and Kunii [2]. They referred to the original definition of L-systems [82] and found them unsuitable to model the complex branching patterns of higher plants. However, their arguments do not extend to parametric L-systems with turtle interpretation. For example, the L-system in Figure 2.6 implements those tree models of Honda [65] in which one of the branching angles is equal to 0, yielding a *monopodial* structure with clearly delineated main and lateral axes (see Section 3.2 for a formal characterization).

*Application of
L-systems*

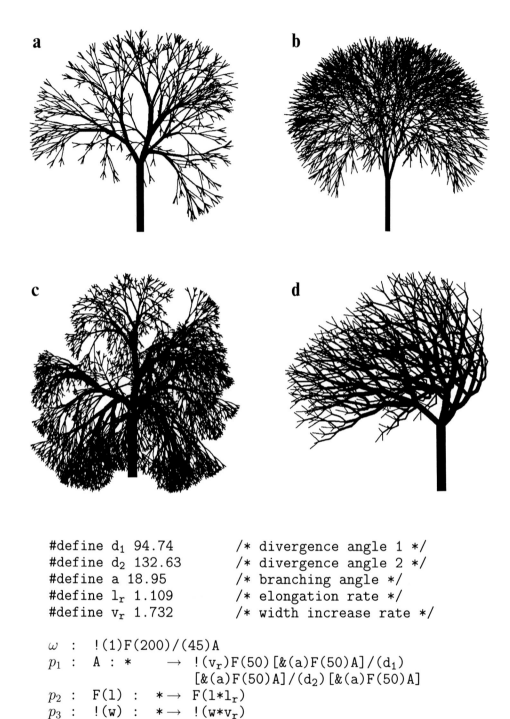

```
#define d₁ 94.74        /* divergence angle 1 */
#define d₂ 132.63       /* divergence angle 2 */
#define a 18.95         /* branching angle */
#define lᵣ 1.109        /* elongation rate */
#define vᵣ 1.732        /* width increase rate */

ω  :   !(1)F(200)/(45)A
p₁ :   A : *      →    !(vᵣ)F(50)[&(a)F(50)A]/(d₁)
                        [&(a)F(50)A]/(d₂)[&(a)F(50)A]
p₂ :   F(l) :   *→    F(l*lᵣ)
p₃ :   !(w) :   *→    !(w*vᵣ)
```

Figure 2.8: Examples of tree-like structures with ternary branching

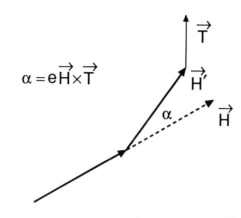

Figure 2.9: Correction α of segment orientation $\vec{H}$ due to tropism $\vec{T}$

Figure	d_1	d_2	a	l_r	$\vec{T}$	e	n
a	94.74	132.63	18.95	1.109	0.00,-1.00,0.00	0.22	6
b	137.50	137.50	18.95	1.109	0.00,-1.00,0.00	0.14	8
c	112.50	157.50	22.50	1.790	-0.02,-1.00,0.00	0.27	8
d	180.00	252.00	36.00	1.070	-0.61,0.77,-0.19	0.40	6

Table 2.3: Constants for the tree structures in Figure 2.8

Figure 2.10: *Medicine Lake* by Musgrave *et al.* [101]

Figure 2.11: A surrealistic elevator

Conclusions   The examples given above demonstrate that the tree models of Honda, as well as their derivatives studied by Aono and Kunii, can be expressed using the formalism of L-systems. In a separate study, Shebell [130] also showed that L-systems can be applied to generate the architectural tree models of Hallé, Oldeman and Tomlinson [58]. These results indicate that L-systems may play an important role as a tool for biologically-correct simulation of tree development and synthesis of realistic tree images. However, the tree-like shapes created so far are rather generic (Figure 2.11), and models of particular tree species, directly based on biological data, are yet to be developed. L-systems have found more applications in the domain of realistic modeling of herbaceous plants, discussed in the next chapter.

Chapter 3

Developmental models of herbaceous plants

The examples of trees presented in the previous chapter introduce L-systems as a plant modeling tool. They also illustrate one of the most striking features of the generative approach to modeling, called *data base amplification* [136]. This term refers to the generation of complex-looking objects from very concise descriptions – in our case, L-systems comprised of small numbers of productions. Yet in spite of the small size, the specification of L-systems is not a trivial task.

In the case of highly self-similar structures, the synthesis methods based on edge rewriting and node rewriting are of assistance, as illustrated by the examples considered in Section 1.10.3. However, a more general approach is needed to model the large variety of developmental patterns and structures found in nature.

The methodology presented in this chapter is based on the simulation of the development of real plants. Thus, a particular form is modeled by capturing the essence of the *developmental process* that leads to this form. This approach has two distinctive features.

Developmental models

- **Emphasis on the space-time relation between plant parts.** In many plants, organs in various stages of development can be observed at the same time. For example, some flowers may still be in the bud stage, others may be fully developed, and still others may have been transformed into fruits. If development is simulated down to the level of individual organs, such *phase effects* are reproduced in a natural way.

- **Inherent capability of growth simulation.** The mathematical model can be used to generate biologically correct images of plants at different ages and to create sequences of images illustrating plant development in time.

The models are constructed under the assumption that organisms control the important aspects of their own development. According to Apter [3, page 44], this simplification must be accepted as a necessary evil, as long as the scope of the mathematical model is limited to an isolated plant. Consequently, this chapter focuses on the modeling and generation of growth sequences of *herbaceous* or non-woody plants, since internal control mechanisms play a predominant role in their development. In contrast, the form of woody plants is determined to a large extent by the environment, competition among branches and trees, and accidents [164].

Herbaceous plants

3.1   Levels of model specification

L-systems can be constructed with a variety of objectives in mind, ranging from a general classification of branching structures to detailed models suitable for image synthesis purposes. Accordingly, the L-systems presented in this chapter are specified at three levels of detail. The most abstract level, called *partial L-systems,* employs the notation of nondeterministic OL-systems to define the realm of possibilities within which structures of a given type may develop. Partial L-systems capture the main traits characterizing structural types, and provide a formal basis for their classification. Control mechanisms that resolve nondeterminism are introduced in the next level, termed *L-system schemata.*[1] The topology of individual plants and temporal aspects of their development are described at this level. Schemata are of particular interest from a biological point of view, as they provide an insight into the mechanisms that control plant development in nature. The geometric aspects are added in *complete L-systems* that include information concerning growth rates of internodes, the values of branching angles, and the appearance of organs. The difference between all three levels is illustrated using models of a single-flower shoot as a running example.

Partial L-systems

L-system schemata

Complete L-systems

3.1.1   Partial L-systems

Consider the development of a shoot which, after a period of vegetative growth, produces a single flower. The partial L-system is given below.

Single-flower shoot

$$
\begin{aligned}
\omega : \quad & a \\
p_1 : \quad & a \rightarrow I[L]a \\
p_2 : \quad & a \rightarrow I[L]A \\
p_3 : \quad & A \rightarrow K
\end{aligned}
\tag{3.1}
$$

The lower-case symbol a represents the *vegetative apex,* while the upper-case A is the *flowering apex,* capable of forming reproductive organs.

---

[1] In the literature, the term "scheme" is also used to denote the class of L-systems with the same alphabet and productions, but with different axioms [62, page 54].

Figure 3.1: Single-flower shoot

A derivation step corresponds to a *plastochron*, defined as the time interval between the production of successive internodes by the apex. At each step apex a has a choice of forming either leaf L, internode I and new apex a (production p_1), or forming the same structures and transforming itself into a flowering apex A (p_2), which subsequently creates flower K (p_3). Once this transformation or *developmental switch* has taken place it cannot be reversed, since there is no rule allowing the transformation of A to a. Examples of strings generated by the L-system specified in equation (3.1) are given below.

$$
\begin{array}{lll}
a & a & a \\
I[L]A & I[L]a & I[L]a \\
I[L]K & I[L]I[L]A & I[L]I[L]a \\
I[L]K & I[L]I[L]K & I[L]I[L]I[L]A \\
& I[L]I[L]K & I[L]I[L]I[L]K \\
& & I[L]I[L]I[L]K \\
\end{array}
$$

A diagrammatic representation of a single-flower inflorescence is shown in Figure 3.1.

3.1.2   Control mechanisms in plants

A partial L-system does not specify the moments in which developmental switches occur. The timing of these switches is specified at the level of L-system schemata, which incorporate mechanisms that control plant development. In biology, these mechanisms are divided into two classes depending on the way information is transferred between modules. The term *lineage* (or *cellular descent*) refers to the transfer of information from an ancestor cell or module to its descendants. In contrast, *interaction* is the mechanism of information exchange between neighboring cells (for example, in the form of nutrients or hormones). Within the formalism of L-systems, lineage mechanisms are represented

L-system schemata

Lineage vs. interaction

by context-free productions found in OL-systems, while the simulation of interaction requires the use of context-sensitive 1L-systems and 2L-systems.[2] Several specific mechanisms are listed below. Although they are described from the modeling perspective, a relation to physiological processes observed in nature can often be found.

Stochastic mechanism

The simplest method for implementing a developmental switch is to use a stochastic L-system. In this case the vegetative apex a has a probability π_1 of staying in the vegetative state, and π_2 of transforming itself into a flowering apex A.

$$
\begin{aligned}
\omega \;&:\; a \\
p_1 \;&:\; a \xrightarrow{\pi_1} I[L]a \\
p_2 \;&:\; a \xrightarrow{\pi_2} I[L]A \\
p_3 \;&:\; A \xrightarrow{1} K
\end{aligned}
$$

The probability distribution (π_1, π_2) is found experimentally, with $\pi_1 + \pi_2 = 1$.

The effect of environment

Many plants change from a vegetative to a flowering state in response to environmental factors such as temperature or the number of daylight hours. Such effects can be modeled using one set of productions (called a *table*) for some number of derivation steps, then replacing it by another set:

Table L-systems

$$
\begin{array}{ll}
\qquad\text{Table 1} & \qquad\text{Table 2} \\
\omega \;:\; a & p_1 \;:\; a \rightarrow I[L]A \\
p_1 \;:\; a \rightarrow I[L]a & p_2 \;:\; A \rightarrow K
\end{array}
$$

The concept of table L-systems (*TOL-systems*) was introduced and formalized by Rozenberg [62, 127]. Note that the use of tables provides only a partial solution to the problem of specifying the switching time, since a control mechanism external to the L-system is needed to select the appropriate table.

Delay mechanism

The delay mechanism operates under the assumption that the apex undergoes a series of state changes that postpone the switch until a particular state is reached.

---

[2] The clarity of this dichotomy is somewhat obscured by parametric OL-systems, which can simulate the operation of context-sensitive L-systems using an infinite set of parameter values.

This is captured by the following L-system in the case of a single-flower shoot.

$$\begin{array}{lll} \omega & : & a_0 \\ p_i & : & a_i \rightarrow I[L]a_{i+1} \quad 0 \leq i \leq n-1 \\ p_n & : & a_n \rightarrow I[L]A \\ p_{n+1} & : & A \rightarrow K \end{array}$$

According to this model, the apex *counts* the leaves it produces. While it may seem strange that a plant would count, it is known that some plant species produce a fixed number of leaves before they start flowering.

Accumulation of components

A developmental mechanism based on the accumulation of components is similar to that of delay, but emphasizes the physiological nature of the counting process. According to this approach, counting is achieved by a monotonic increase or decrease in the concentration of certain cell components. This process can be captured by the following parametric L-system:

$$\begin{array}{lllll} \omega & : & a(0) \\ p_1 & : & a(c) & : & c < C \rightarrow I[L]a(c + \Delta c) \\ p_2 & : & a(c) & : & c \geq C \rightarrow I[L]A \\ p_3 & : & A & : & * \quad \rightarrow K \end{array} \tag{3.2}$$

The parameter c indicates current concentration of the controlling components in the apex a. In each derivation step, this concentration is increased by a constant Δc. The developmental switch occurs when the concentration reaches the threshold value C.

Development controlled by a signal

In many plants, the switch from a vegetative to a flowering state is caused by a flower-inducing signal transported from the basal leaves towards the apex. The time of signal initiation is determined using one of the previously described methods, for example by counting. A sample L-system is given below.

$$\begin{array}{lllll} \omega & : & D(1)a(1) \\ p_1 & : & & a(i) & : & i < m \rightarrow a(i+1) \\ p_2 & : & & a(i) & : & i = m \rightarrow I[L]a(1) \\ p_3 & : & & D(i) & : & i < d \rightarrow D(i+1) \\ p_4 & : & & D(i) & : & i = d \rightarrow S(1) \\ p_5 & : & & S(i) & : & i < u \rightarrow S(i+1) \\ p_6 & : & & S(i) & : & i = u \rightarrow \varepsilon \\ p_7 & : & S(i) < I & : & i = u \rightarrow IS(1) \\ p_8 & : & S(i) < a(j) & : & * \quad \rightarrow I[L]A \\ p_9 & : & & A & : & * \quad \rightarrow K \end{array}$$

The apex a produces internodes I and leaves L on the main axis (p_2). The time between the production of two consecutive segments, or the plastochron of the main axis, is equal to m derivation steps (p_1). After a delay of d steps (p_3), a signal S is sent from the plant base towards the apices (p_4). This signal is transported along the main axis with a delay of u steps per internode I (p_5,p_7). Production p_6 removes the signal from a node after it has been transported along the structure (ε stands for the empty string). When the signal reaches the apex, a is transformed into flowering state A (p_8) which yields flower K (p_9). Note that the signal has to propagate faster than one node per plastochron $(u < m)$, otherwise it would not be able to catch up with the apex. The above processes are illustrated by the following developmental sequence, for $d = 4$, $m = 2$ and $u = 1$.

$$
\begin{aligned}
&D(1)a(1)\\
&D(2)a(2)\\
&D(3)I[L]a(1)\\
&D(4)I[L]a(2)\\
&S(1)I[L]I[L]a(1)\\
&IS(1)[L]I[L]a(2)\\
&I[L]IS(1)[L]I[L]a(1)\\
&I[L]I[L]IS(1)[L]a(2)\\
&I[L]I[L]I[L]A\\
&I[L]I[L]I[L]K
\end{aligned}
$$

Although the above model may appear unnecessarily complicated, signals are indispensable in the simulation of complex flowering sequences discussed later.

3.1.3   Complete models

The L-systems considered so far are not directly suitable for image synthesis purposes. To this end, they must be completed with geometric information. The relation between an L-system scheme and a corresponding complete L-system is discussed using the model of crocuses shown in Figure 3.2 as an example.

Crocus

The development is controlled using a delay expressed as an accumulation mechanism (equation (3.2)). In contrast to L-system schemes in which symbols represent module types, the L-system in Figure 3.2 is specified in terms of turtle symbols. Production p_1 describes the creation of successive internodes F and leaves L by the vegetative apex a. The leaves branch from the stem at an angle of 30° and spiral around the main axis with a divergence angle equal to 137.5° (see Chapter 4). Productions p_2 and p_3 describe the developmental switch and the creation of flower K taking place respectively in steps T_a and T_{a+1}. Productions p_4 and p_5 capture the development of leaves and flowers until they reach their final shapes T_L and T_K steps after creation. For each

```
#define Tₐ 7              /* developmental switch time */
#define T_L 9             /* leaf growth limit */
#define T_K 5             /* flower growth limit */
#include L(0),L(1),...,L(T_L)  /* leaf shapes */
#include K(0),K(1),...,K(T_K)  /* flower shapes */

ω  :  a(1)
p₁ :  a(t)  :  t<Tₐ   →   F(1)[&(30)∼L(0)]/(137.5)a(t+1)
p₂ :  a(t)  :  t=Tₐ   →   F(20)A
p₃ :  A     :  *      →   ∼K(0)
p₄ :  L(t)  :  t<T_L  →   L(t+1)
p₅ :  K(t)  :  t<T_K  →   K(t+1)
p₆ :  F(l)  :  l<2    →   F(l+0.2)
```

Figure 3.2: Crocuses

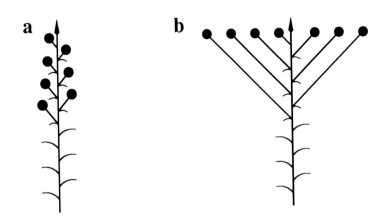

Figure 3.3: Open racemes: (a) elongated form, (b) planar form

Figure 3.4: Lily-of-the-valley

The flowering sequence in open racemes is always *acropetal* (from base to top). This can be observed after substituting production p_3 in the L-system in equation (3.3) with productions p_3' and p_4, which use indexed symbols K_i to denote subsequent stages of flower development.

$$
\begin{array}{rll}
p_3': & A & \rightarrow \quad I[K_0]A \\
p_4: & K_i & \rightarrow \quad K_{i+1}, \qquad i \geq 0
\end{array}
$$

The indexed notation $K_i \rightarrow K_{i+1}$ stands for a (potentially infinite) set of productions $K_0 \rightarrow K_1, K_1 \rightarrow K_2, K_2 \rightarrow K_3, \ldots$. The developmental sequence begins as follows:

$$
\begin{array}{l}
A \\
I[IK_0]A \\
I[IK_1]I[IK_0]A \\
I[IK_2]I[IK_1]I[IK_0]A \\
I[IK_3]I[IK_2]I[IK_1]I[IK_0]A \\
\cdots
\end{array}
$$

At each developmental stage the inflorescence contains a sequence of flowers of different ages. The flowers newly created by the apex are delayed in their development with respect to the older ones situated at the stem base. Graphically, this effect is illustrated by the model of a lily-of-the-valley shown in Figure 3.4. The following quotation from d'Arcy Thompson [143] applies:

Lily-of-the-valley

> A flowering spray of lily-of-the-valley exemplifies a growth-gradient, after a simple fashion of its own. Along the stalk the growth-rate falls away; the florets are of descending age, from flower to bud; their graded differences of age lead to an exquisite gradation of size and form; the time-interval between one and another, or the "space-time relation" between them all, gives a peculiar quality – we may call it phase-beauty – to the whole.

Another example of "phase beauty" can be seen in the shoot of shepherd's purse (*Capsella bursa-pastoris*) shown in Figure 3.5. Productions p_1, p_2 and p_3 describe the activities of the apex in the vegetative and flowering states, in accordance with the L-system in equation (3.3). The developmental switch is implemented using a delay mechanism. Productions p_4 and p_5 capture the linear elongation of internodes in time, while p_6 and p_7 describe the gradual increase of the angle at which the flower stalks branch from the main stem. Productions p_8, p_9 and p_{11} specify the shapes of leaves L, flower petals K and fruits X using developmental surface models discussed in Section 5.2. Production p_{10} controls the flowering time. Symbol % in the successor of production p_{11} simulates the fall of petals by cutting them off the structure at the time of fruit formation. The default value of the angle increment corresponding to the symbol + with no parameter is 18°.

Capsella

$$\omega \quad : \quad \texttt{I(9)a(13)}$$
$$p_1 \quad : \quad \texttt{a(t)} \quad : \quad \texttt{t>0} \quad \rightarrow \quad \texttt{[\&(70)L]/(137.5)I(10)a(t-1)}$$
$$p_2 \quad : \quad \texttt{a(t)} \quad : \quad \texttt{t=0} \quad \rightarrow \quad \texttt{[\&(70)L]/(137.5)I(10)A}$$
$$p_3 \quad : \quad \texttt{A} \quad : \quad \texttt{*} \quad \rightarrow \quad \texttt{[\&(18)u(4)FFI(10)I(5)X(5)KKKK]}$$
$$\phantom{p_3 \quad : \quad \texttt{A} \quad : \quad \texttt{*} \quad \rightarrow \quad} \texttt{/(137.5)I(8)A}$$
$$p_4 \quad : \quad \texttt{I(t)} \quad : \quad \texttt{t>0} \quad \rightarrow \quad \texttt{FI(t-1)}$$
$$p_5 \quad : \quad \texttt{I(t)} \quad : \quad \texttt{t=0} \quad \rightarrow \quad \texttt{F}$$
$$p_6 \quad : \quad \texttt{u(t)} \quad : \quad \texttt{t>0} \quad \rightarrow \quad \texttt{\&(9)u(t-1)}$$
$$p_7 \quad : \quad \texttt{u(t)} \quad : \quad \texttt{t=0} \quad \rightarrow \quad \texttt{\&(9)}$$
$$p_8 \quad : \quad \texttt{L} \quad : \quad \texttt{*} \quad \rightarrow \quad \texttt{[\{.-FI(7)+FI(7)+FI(7)\}]}$$
$$\phantom{p_8 \quad : \quad \texttt{L} \quad : \quad \texttt{*} \quad \rightarrow \quad} \texttt{[\{.+FI(7)-FI(7)-FI(7)\}]}$$
$$p_9 \quad : \quad \texttt{K} \quad : \quad \texttt{*} \quad \rightarrow \quad \texttt{[\&\{.+FI(2)--FI(2)\}]}$$
$$\phantom{p_9 \quad : \quad \texttt{K} \quad : \quad \texttt{*} \quad \rightarrow \quad} \texttt{[\&\{.-FI(2)++FI(2)\}]/(90)}$$
$$p_{10} \quad : \quad \texttt{X(t)} \quad : \quad \texttt{t>0} \quad \rightarrow \quad \texttt{X(t-1)}$$
$$p_{11} \quad : \quad \texttt{X(t)} \quad : \quad \texttt{t=0} \quad \rightarrow \quad \texttt{\textbackslash(50)[[-GGGG++[GGG[++G\{.].].].}$$
$$\phantom{p_{11} \quad : \quad \texttt{X(t)} \quad : \quad \texttt{t=0} \quad \rightarrow \quad} \texttt{++GGGG.--GGG.--G.\}]\%}$$

Figure 3.5: Development of *Capsella bursa-pastoris*. Every fourth derivation step is shown.

Figure 3.6: Apple twig

Simple raceme (closed)

The inflorescence of an apple tree (Figure 3.6) provides an example of a closed raceme. In this case, the main apex eventually terminates its development and produces a terminal flower (Figure 3.7). The corresponding partial L-system is given below.

$$
\begin{aligned}
\omega &: \quad a \\
p_1 &: \quad a \rightarrow I[L]a \\
p_2 &: \quad a \rightarrow I[L]A \\
p_3 &: \quad A \rightarrow I[K]A \\
p_4 &: \quad A \rightarrow K
\end{aligned}
$$

Developmental switches are associated with two symbols, a and A. Thus, in order to obtain an L-system scheme it is necessary to specify how both of these switches will be controlled.

The flowering sequence is usually acropetal but could also be basipetal, *i.e.*, progressing downward after the formation of the terminal flower on the main axis. In the latter case a basipetal signal, as discussed in Section 1.8, can be applied to induce the transformation of dormant flower buds into flowers.

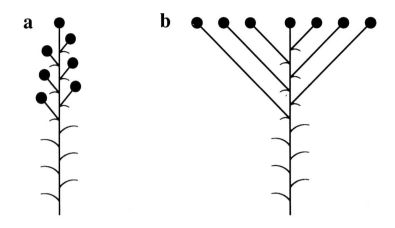

Figure 3.7: Closed racemes: (a) elongated form, (b) planar form

Compound raceme (open dibotryoid)

Racemes can also occur on complex branching structures. The simplest of these inflorescences is one with open racemes on the first order branches as well as on the main axis (Figure 3.8a). This two-level compound structure (thus *dibotryoid*) is described by the following partial L-system.

$$
\begin{aligned}
\omega &: \quad a \\
p_1 &: \quad a \rightarrow I[L]a \\
p_2 &: \quad a \rightarrow I[L]A \\
p_3 &: \quad A \rightarrow I[L][b]A \\
p_4 &: \quad A \rightarrow I[L][b]B \\
p_5 &: \quad b \rightarrow I[L]b \\
p_6 &: \quad b \rightarrow I[L]B \\
p_7 &: \quad B \rightarrow I[K]B
\end{aligned}
\tag{3.4}
$$

Three developmental transformations are necessary: the first for the change from leaf to branch creation along the main axis (production p_2), the second for the switch from branching to lateral flower creation on the main axis (p_4), and the third for the transition from leaf to lateral flower formation along the first-order branches (p_6). Each branch is subtended by a leaf, which is why productions p_3 and p_4 specify two appendages L and b. Branches with flowers K need not have subtending leaves, which is reflected in production p_7.

Single-signal model

Within each component raceme, the flowering sequence is always acropetal, but the timing of switches has a crucial impact on the overall flowering sequence and appearance of the plant. For example, let us assume that the switch from leaf to branch production is controlled by a delay, while the remaining two switches are caused by an acropetal flower-inducing signal (representing the hormone *florigen*). Such a development is captured by L-system 3.1 (see below). Initially, the veg-

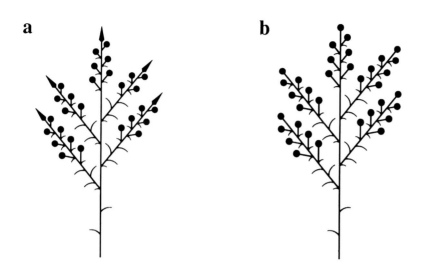

a   **b**

Figure 3.8: Dibotryoids: (a) open, (b) closed

```
#define d 13      /* delay for sending florigen */
#define e 3       /* delay for creating branches */
#define m 2       /* plastochron - main axis */
#define n 3       /* plastochron - lateral axis */
#define u 1       /* signal delay - main axis */
#define v 1       /* signal delay - lateral axis */
```

$$\omega \quad : S(0)a(1,0)$$

p_1 :   a(t,c) : t<m   → a(t+1,c)

p_2 :   a(t,c) : (t=m)&(c<e) → I(0,u)[L]a(1,c+1)

p_3 :   a(t,c) : (t=m)&(c=e) → I(0,u)[L][b(1)]a(1,c)

p_4 :   b(t)   : t<n   → b(t+1)

p_5 :   b(t)   : t=n   → I(0,v)[L]b(1)

p_6 :   S(t)   : *   → S(t+1)

p_7 :   S(t) < I(i,j) : t=d   → I(1,j)

p_8 :   I(i,j) : (0<i)&(i<j) → I(i+1,j)

p_9 : I(i,j) < I(k,l) : (i=j)&(k=0) → I(1,l)

p_{10}: I(i,j) < a(k,l) : i>0   → I[L][b(1)]B

p_{11}: I(i,j) < b(k)   : i>0   → I[L]B

p_{12}:   B   : *   → I[K]B

L-system 3.1: A model of dibotryoids

etative apex a creates internodes I and leaves L with plastochron m (productions p_1 and p_2). After the creation of e leaves a developmental switch occurs, and apex a starts creating branches with the same plastochron (p_3). The change of state is indicated by the value of the second parameter in the module $a(t, c)$, which is now equal to e. The lateral apices b create internodes and leaves with plastochron n (p_4 and p_5). After a delay of d steps from the beginning of the simulation (p_6), the flowering signal is introduced to the basal internode (p_7), as indicated by a non-zero value of the first parameter in the module $I(i, j)$. The signal is passed along an axis at the rate of j steps per internode (p_8 and p_9), where $j = u$ for the main axis and $j = v$ for the lateral axes. These rates are assigned to internodes by productions p_2, p_3 and p_5. When the signal reaches an apex (either a or b), the apex is transformed into flowering state B (p_{10} and p_{11}). From then on, new flowers K are produced in each derivation step (p_{12}).

Model analysis   In order to analyze the plant structure and flowering sequence resulting from the above development, let T_k denote the time at which apex b of the k-th lateral axis is transformed into the flowering state, and l_k denote the length of this axis (expressed as the number of internodes) at the transformation time. It is assumed here that the first e leaves count as lateral axes, thus $k > e$. Since it takes km time units to produce k internodes along the main axis and $l_k n$ time units to produce l_k internodes on the lateral axis, we obtain:

$$T_k = km + l_k n$$

On the other hand, the transformation occurs when the signal reaches the apex. The signal is sent d time units after the development starts. It uses ku time units to travel through k zero-order internodes and $l_k v$ time units to travel through l_k first-order internodes:

$$T_k = d + ku + l_k v$$

Solving the above system of equations for l_k and T_k (and ignoring for simplicity some inaccuracy due to the fact that this system does not guarantee integer solutions), we obtain:

$$T_k = k\frac{un - vm}{n - v} + d\frac{n}{n - v}$$

$$l_k = -k\frac{m - u}{n - v} + \frac{d}{n - v}$$

In order to analyze the above solutions, let us first notice that the signal transportation delay v must be less than the plastochron of the lateral axes n, otherwise the signal would never reach the lateral apices. Under this assumption, the sign of the expression $\Delta = un - vm$ determines the overall flowering sequence, which is acropetal for $\Delta > 0$ (Figure 3.9) and basipetal for $\Delta < 0$ (Figure 3.10). If $\Delta = 0$, all flowering switches

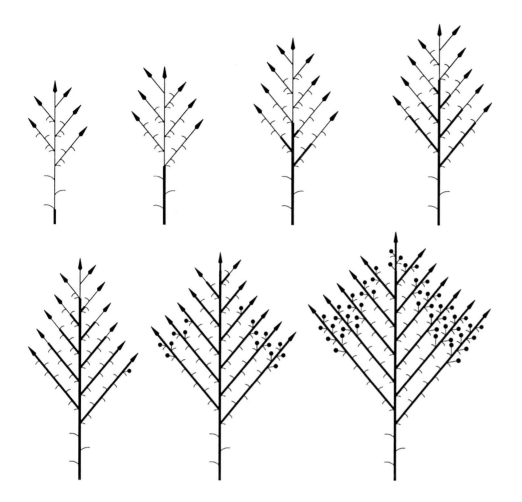

Figure 3.9: An acropetal flowering sequence in an open dibotryoid: $m = 2$, $n = 3$, $u = v = 1$, $\Delta = 0.5$; derivation lengths: $15 - 18 - 21 - 24 - 27 - 30 - 33$

occur simultaneously. The sign of the expression $m - u$ determines whether the vegetative part of the shoot is more developed at the base $(m - u < 0)$ or near the top of the structure $(m - u > 0)$. Figure 3.11 shows a model of a member of the mint family that exhibits a basipetal flowering sequence.

Compound racemes (closed dibotryoids)

This inflorescence type differs from the previous one only in that each branch, including the main axis, bears a terminal flower (Figure 3.8b). A partial L-system can be obtained from that of equation (3.4) by adding one more production:

$$p_8 : B \to K$$

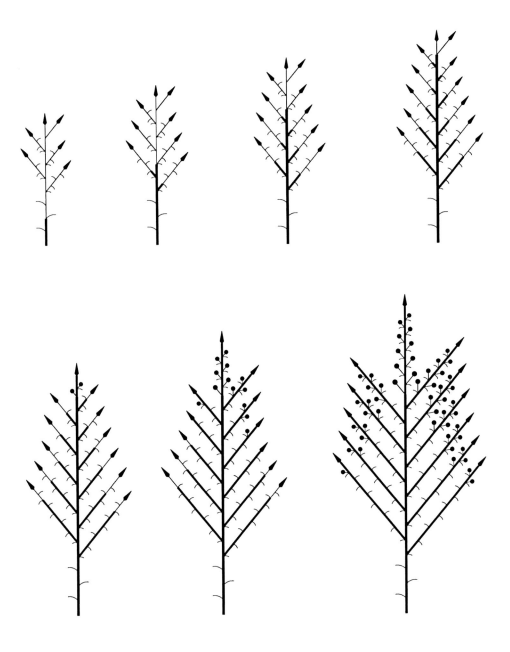

Figure 3.10: A basipetal flowering sequence in an open dibotryoid: $m = 2$, $n = 5$, $u = 1$, $v = 3$, $\Delta = -0.5$; derivation lengths: $16 - 20 - 24 - 28 - 32 - 36 - 40$

Figure 3.11: A mint

Compound raceme (closed tribotryoid)

Racemic inflorescences can be compounded to a higher number of levels. The following is a partial L-system for a closed tribotryoid inflorescence, where closed racemes occur on second-order branches as well as on the terminal portions of first-order branches and of the main axis (Figure 3.12). The developmental process involves six developmental transformations.

$$\omega : \quad a$$
$$p_1 : \quad a \rightarrow I[L]a$$
$$p_2 : \quad a \rightarrow I[L]A$$
$$p_3 : \quad A \rightarrow I[L][b]A$$
$$p_4 : \quad A \rightarrow I[L][b]B$$
$$p_5 : \quad b \rightarrow I[L]b$$
$$p_6 : \quad b \rightarrow I[L]B$$

$$p_7 : \quad B \rightarrow I[L][c]B$$
$$p_8 : \quad B \rightarrow I[L][c]C$$
$$p_9 : \quad c \rightarrow I[L]c$$
$$p_{10} : \quad c \rightarrow I[L]C$$
$$p_{11} : \quad C \rightarrow I[K]C$$
$$p_{12} : \quad C \rightarrow K$$

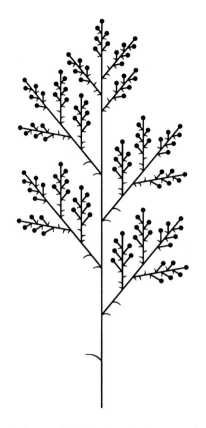

Figure 3.12: Closed tribotryoid

3.3.2   Symodial inflorescences

Simple cymes (open)

In racemes, the apex of the main axis produces lateral branches and
continues to grow. In contrast, the apex of the main axis in *cymes* turns
into a flower shortly after a few lateral branches have been initiated.
Their apices turn into flowers as well, and second-order branches take
over. In time, branches of higher and higher order are produced. Thus,
the basic structure of a cymose inflorescence is captured by the partial
L-system:

$$
\begin{aligned}
\omega &: \quad a \\
p_1 &: \quad a \rightarrow I[L]a \\
p_2 &: \quad a \rightarrow I[L]A \\
p_3 &: \quad A \rightarrow I[A]K
\end{aligned}
\tag{3.5}
$$

As in the open raceme, there is a single symbol with alternative rules
which specify that the vegetative apex a may change into a flower-
producing apex A. Any one of the previously discussed mechanisms
is available for timing this decision. Figure 3.13a shows an open cyme
with branches curving in a spiral fashion, while Figure 3.13b shows one

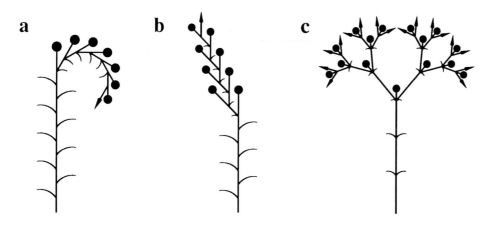

Figure 3.13: Open cymes: (a) spiral form, (b) zig-zag form, (c) double

with a zig-zag branching form.

Double cymes (open)

Frequently, not one but two lateral apices are produced under each terminal apex as in Figure 3.13c. In this case the partial L-system is:

$$
\begin{aligned}
\omega &: \quad a \\
p_1 &: \quad a \rightarrow I[L]a \\
p_2 &: \quad a \rightarrow I[L]A \\
p_3 &: \quad A \rightarrow I[A][A]K
\end{aligned} \tag{3.6}
$$

The two continuing lateral apices may develop at approximately equal rates (with the same plastochron) or with different rates, giving rise to asymmetric inflorescences. For example, the following L-system scheme describes the development of rose campion (*Lychnis coronaria*) as analyzed by Robinson [126]:

$$
\begin{aligned}
\omega &: \quad A_7 \\
p_1 &: \quad A_7 \rightarrow I[A_0][A_4]IK_0 \\
p_2 &: \quad A_i \rightarrow A_{i+1}, \qquad 0 \leq i < 7 \\
p_3 &: \quad K_i \rightarrow K_{i+1}, \qquad i \geq 0
\end{aligned}
$$

Lychnis

Production p_1 shows that at their creation time, the lateral apices have different states A_0 and A_4. Consequently, the first apex requires eight derivation steps to produce a flower and new branches, while the second requires only four steps. Each flower undergoes a sequence of changes, progressing from the bud stage to an open flower to a fruit. This developmental sequence is illustrated in Figure 3.14. According to production p_1, the lateral apices branch at an angle of 45° and lie in a plane perpendicular to that defined by the mother axis and its sibling. Production p_3 describes the linear elongation of internodes, while p_4

```
#include L(0),L(1),...   /* leaf shapes */
#include K(0),K(1),...   /* flower shapes */

ω  : A(7)
p₁ : A(t) : t=7 → FI(20)[&(60)∼L(0)]/(90)[&(45)A(0)]/(90)
                  [&(60)∼L(0)]/(90)[&(45)A(4)]FI(10)∼K(0)
p₂ : A(t) : t<7 → A(t+1)
p₃ : I(t) : t>0 → FFI(t-1)
p₄ : L(t) : *   → L(t+1)
p₅ : K(t) : *   → K(t+1)
```

Figure 3.14: Development of *Lychnis coronaria*

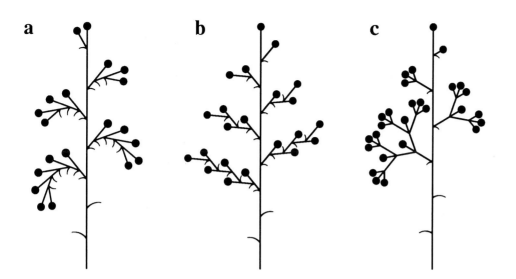

Figure 3.15: Thyrsus: (a) spiral form, (b) zig-zag form, (c) double

and p_5 capture the development of leaves and flowers over time. It is interesting to note that at different developmental stages there are some open flowers that have a relatively uniform distribution over the entire plant structure. This is advantageous to the plant since it increases the time span over which seeds will be produced.

Cymes (closed)

Sympodial inflorescences that produce a terminal flower at some point during their development are called *closed cymes*. They result from the addition of production

$$p_4 : A \rightarrow K$$

to the L-systems specified in (3.5) and (3.6), which define open single and open double cymes.

Thyrsus (closed)

A *thyrsus* is an inflorescence with branches of cymes borne on a monopodially branching axis. Thus, it represents a mixed sympodial and monopodial organization. Depending on the orientation of the flowers, a distinction between a thyrsus with cymes in a spiral form and in a zig-zag form can be made (Figure 3.15, a and b). Both of these types are described by the following partial L-system:

$$
\begin{array}{rl}
\omega : & a \\
p_1 : & a \to I[L]a \\
p_2 : & a \to I[L]A \\
p_3 : & A \to I[L][B]A \\
p_4 : & A \to K \\
p_5 : & B \to I[B]K \\
p_6 : & B \to K
\end{array}
$$

In addition, a thyrsus may have double cymes (Figure 3.15c). In the closed structure there are three developmental transformations. The first represents the change from vegetative to flowering development on the main axis (production p_2). The second is necessary for the closure of the main axis with a terminal flower (p_4). Both switches are related to the monopodial development of the main axis. The third transformation is responsible for the formation of the flowers that terminate the development of the sympodial structures (p_6).

3.3.3   Polypodial inflorescences

Panicle

The term *polypodial* is not used in the botanical literature but is coined here to draw attention to the type of branching that represents continuing development of the main axis as well as of the lateral apices of a branch. The corresponding inflorescence type is usually called a *panicle*. The presence of two continuing apices at each new node is expressed by the following production:

$$
A \to I[L][A]A
$$

Since there can be nodes near the base of the plant that do not bear branches, the usual initial rules are included to model the transition from a purely vegetative to a flowering state. The resulting partial L-system is:

$$
\begin{array}{rl}
\omega : & a \\
p_1 : & a \to I[L]a \\
p_2 : & a \to I[L]A \\
p_3 : & A \to I[L][A]A \\
p_4 : & A \to K
\end{array}
$$

An example of a paniculate structure is shown in Figure 3.16. Note the presence of higher order branching and the lack of terminal racemes. Due to the repetitive application of production p_3 at various levels of branching, the resulting structure is highly self-similar. The model includes only two types of developmental transformations: the switch

Figure 3.16: Panicle (elongated form)

from purely vegetative growth to the formation of the branching structure (production p_2), and the creation of terminal flowers (p_4). The timing of the last production determines the flowering sequence of the plant. Two possible control mechanisms will be examined in detail, using developmental models of the branching part of wall lettuce (*Mycelis muralis*) as examples.

The development of *Mycelis* is difficult to model for two reasons. *Mycelis* First, the plant exhibits a basipetal flowering sequence, which means that flowering starts at the top of the plant and proceeds downwards. Secondly, at some developmental stages the plant has an *acrotonic* structure, where the upper branches are more developed than the lower ones. Both phenomena are in a sense counter-intuitive, since it would seem that the older branches situated near the plant base should start growing and producing flowers before the younger ones at the plant top. To explain these effects, several models were proposed and formally analyzed by Janssen and Lindenmayer [77]. Their *model II* is restated here as parametric L-system 3.2.

The axiom consists of three components. Modules F and $A(0)$ rep- *Model II* resent the initial segment and the apex of the main axis. Module $I(20)$ is the source of a signal representing florigen. In nature, florigen is sent towards the apex by leaves located at the plant base, which is not included in this model.

The developmental process consists of two phases that take place along the main axis and are repeated recursively in branches of higher orders. First, the main axis is formed in a process of subapical growth

```
#include O                    /* flower shape specification */
#ignore / + ~ O

ω  :   I(20)FA(0)
p₁ :       S < A(t)        : *   → T(0)~O
p₂ :           A(t)        : t>0 → A(t-1)
p₃ :           A(t)        : t=0 → [+(30)G]F/(180)A(2)
p₄ :       S < F           : *   → FS
p₅ :           F > T(c)    : *   → T(c+1)FU(c-1)
p₆ : U(c) < G              : *   → I(c)FA(2)
p₇ :           I(c)        : c>0 → I(c-1)
p₈ :           I(c)        : c=0 → S
p₉ :           S           : *   → ε
p₁₀:           T(c)        : *   → ε
```

L-system 3.2: *Mycelis muralis* – Model II

specified by production p_3. The apex produces consecutive segments
F at the rate of one segment every three derivation steps (the delay is
controlled by production p_2), and initiates branches G positioned at an
angle of 30° with respect to the main axis. The symbol G is interpreted
here in the same way as F. At this stage, the branches do not develop
further, which simulates the effect of *apical dominance* or the inhibition
of branch development during the active production of new branches
by the apex.

After a delay of 20 derivation steps, counted using production p_7,
an acropetal flower-inducing signal S is sent by production p_8. Production p_4 transports S across the segments at the rate of one internode
per step. Since new internodes are produced by the apex at a three
times slower rate, the signal eventually reaches the apex. At this point,
the second developmental phase begins. Production p_1 transforms apex
$A(t)$ into a bud O. Further branch production is stopped and a signal
$T(c)$ is sent towards the base in order to enable the development of
lateral branches. Parameter c is incremented by production p_5 each
time signal $T(c)$ traverses an internode. Subsequently, production p_6
introduces the value of parameter c into the corresponding branches,
using module $U(c)$ as a carrier. The successor of production p_6 has
the same format as the axiom, thus module $I(c)$ determines the delay
between the initiation of branch development and time signal S, sent
to terminate further internode creation. This delay c is smallest for
the top branches and increases towards the plant base. Consequently,
parameter c can be interpreted as the *growth potential* of the branches,
allowing lower branches to grow longer than the higher ones. On the
other hand, the development of the upper branches starts sooner, thus
in some stages they will be more developed than the lower ones, and
the flowering sequence will progress downwards, corresponding to ob-

```
#include K               /* flower shape specification */
#consider M S T V

ω  :   I(20)FA(0)
p₁ :   S < A(t)  : *    → TV K
p₂ :   V < A(t)  : *    → TV K
p₃ :       A(t)  : t>0  → A(t-1)
p₄ :       A(t)  : t=0  → M[+(30)G]F/(180)A(2)
p₅ :   S < M     : *    → S
p₆ :       S > T : *    → T
p₇ :   T < G     : *    → FA(2)
p₈ :   V < M     : *    → S
p₉ :       T > V : *    → W
p₁₀:       W     : *    → V
p₁₁:       I(t)  : t>0  → I(t-1)
p₁₂:       I(t)  : t=0  → S
```

L-system 3.3: *Mycelis muralis* – Model III

servations of the real plant [77].

A diagrammatic developmental sequence of *Mycelis muralis* simulated using L-system 3.2 is shown in Figure 3.17. Initially, the segments are shown as bright green. The passage of florigen S turns them purple, and the lifting of apical dominance changes their color to dark green. Figure 3.18 represents a three-dimensional rendering of the same model. The three-dimensional structure differs from the two-dimensional diagram only in details. The angle value associated with the module "/" in production p_3 has been changed to 137.5°, resulting in a spiral arrangement of lateral branches around the mother axis. The leaves subtending branches have been included in the model, and flowers have been assumed to undergo a series of changes from bud to open flower to fruit.

Another developmental model of *Mycelis*, referred to here as *model III*, is given by L-system 3.3. The initial phases of development are the same as in model II. First, apex A creates the main axis and initiates lateral branches (productions p_3 and p_4). Symbol M in the successor of production p_4 marks consecutive branching points. After a delay of 20 steps (ω) counted by production p_{11}, flowering signal S is generated at the inflorescence base (p_{12}) and sent up along the main axis (p_5). Upon reaching the apex, S induces its transformation into a terminal flower K, and initiates two basipetal signals T and V (p_1). The basipetal signals also can be initiated by production p_2, which is needed for the proper timing of signals in the topmost lateral branch. Signal T propagates basipetally at the rate of one internode per derivation step (p_6) and lifts apical dominance, thus allowing the lateral branches to grow (p_7). The presence of the second basipetal signal V is the distinctive

Model III

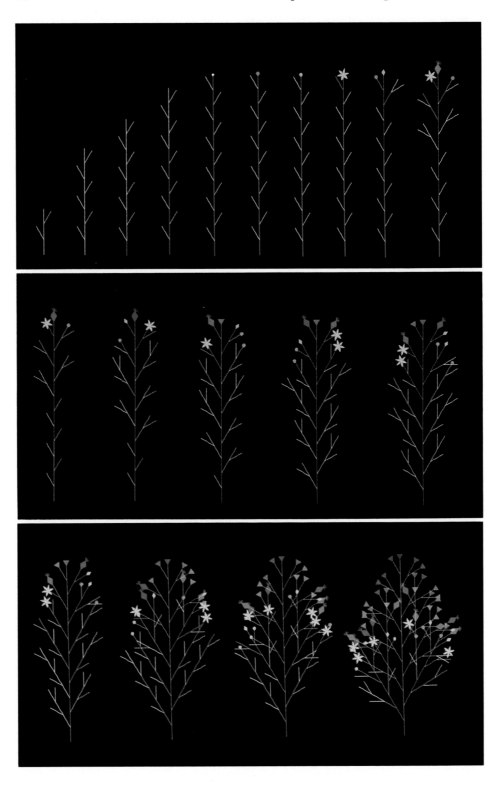

Figure 3.17: Development of *Mycelis muralis*

Figure 3.18: A three-dimensional rendering of the *Mycelis* model

feature of model III. Its role is to enable the formation of flowers on the lateral branches by generating the flowering signal S at their bases (p_8). Since signal V propagates down the main axis at the rate of one internode per two derivation steps (p_9, p_{10}), the interval between the lifting of apical dominance by signal T and induction of flowering signal S by signal V increases linearly towards the inflorescence base. This allows the lower branches to grow longer than the upper ones, resulting in a structure that is more developed near the base than near the apex in later developmental stages.

This entire control process repeats recursively for each axis: its apex is transformed into a flower by signal S, the growth of lateral axes is successively enabled by signal T, and the second basipetal signal V is sent to induce the flowering signal S in the next-order axes. Consequently, a basipetal flowering sequence is observed along all axes of the panicle.

Model II controls the flowering on lateral branches using growth potential c accumulated by signal T on its way down, while model III employs the time interval between signals T and V for the same purpose. Since both models produce identical developmental sequences, it is not possible to decide which one is more faithful to nature without gathering additional data related to plant physiology. Nevertheless, the models clearly indicate that the flowering sequence of *Mycelis* cannot be explained simply in terms of two commonly recognized mechanisms,

Biological relevance

Figure 3.21: The garden of L

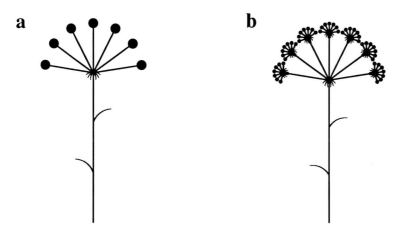

Figure 3.22: Umbels: (a) simple, (b) compound

Umbel

An *umbel* is characterized by more than two internodes attached to a single node, resulting in a typical umbrella-like shape. In a simple umbel there are flowers at the ends of the lateral internodes (Figure 3.22a), while in compound umbels the branching pattern is repeated recursively a certain number of times (Figure 3.22b). The partial L-system for a simple umbel is

$$\begin{aligned}
\omega &: \quad A \\
p_1 &: \quad A \to I[IK]^n
\end{aligned}$$

and for a compound umbel of recursion depth two is

$$\begin{aligned}
\omega &: \quad A \\
p_1 &: \quad A \to I[IB]^k B \\
p_2 &: \quad B \to I[IC]^l C \\
p_3 &: \quad C \to I[IK]^m
\end{aligned}$$

This type of inflorescence is commonly found in the family Umbelliferae. For example, Figure 3.23 presents a model of a wild carrot. Note that the size of leaves decreases towards the top of the plant, producing a phase effect similar to that observed in simple racemes. In contrast, the most developed inflorescence is placed at the top of the plant, indicating developmental control by a hormone similar to that observed in mints (Figure 3.11).

Wild carrot

Spike

An elongated raceme with closely packed flowers is called a *spike*. Many grasses and sedges have this kind of inflorescence (Figure 3.24a). See Figure 4.17 (page 117) for a realistic model.

Figure 3.23: Wild carrot

Spadix

A fleshy elongated raceme is called a *spadix*, and is frequently found in the family Araceae (Figure 3.24b).

Capitulum

A fleshy spherical or disk-shaped raceme is called a *capitulum* or "head." The sunflower head is an inflorescence of this kind, the oldest flowers being at the margin and the youngest at the center (Figure 3.24c). Members of the family Compositae commonly have this type of structure. One characteristic feature is the spatial arrangement of components, such as flowers or seeds, which form early discernible spiral patterns. A detailed description of these patterns is presented in the next chapter.

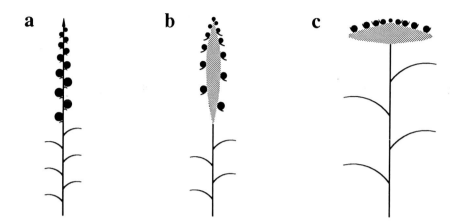

Figure 3.24: Modified racemes: (a) spike, (b) spadix, (c) capitulum

Chapter 4

Phyllotaxis

The regular arrangement of lateral organs (leaves on a stem, scales on a cone axis, florets in a composite flower head) is an important aspect of plant form, known as *phyllotaxis*. The extensive literature generated by biologists' and mathematicians' interest in phyllotaxis is reviewed by Erickson [36] and Jean [78]. The proposed models range widely from purely geometric descriptions (for example, Coxeter [17]) to complex physiological hypotheses tested by computer simulations (Hellendoorn and Lindenmayer [59], Veen and Lindenmayer [151], Young [163]). This chapter presents two models suitable for the synthesis of realistic images of flowers and fruits that exhibit spiral phyllotactic patterns.

Both models relate phyllotaxis to packing problems. The first operates in a plane and was originally proposed by Vogel [154] to describe the structure of a sunflower head. A further detailed analysis was given by Ridley [124, 125]. The second model reduces phyllotaxis to the problem of packing circles on the surface of a cylinder. Its analysis was presented by van Iterson [75] and reviewed extensively by Erickson [36].

The area of phyllotaxis is dominated by intriguing mathematical relationships. One of them is the "remarkable fact that the numbers of spirals which can be traced through a phyllotactic pattern are predominantly integers of the Fibonacci sequence" [36, p. 54]. For example, Coxeter [17] notes that the pineapple displays eight rows of scales sloping to the left and thirteen rows sloping to the right. Furthermore, it is known that the ratios of consecutive Fibonacci numbers F_{k+1}/F_k converge towards the golden mean $\tau = (\sqrt{5}+1)/2$. The *Fibonacci angle* $360°\tau^{-2}$, approximately equal to 137.5°, is the key to the first model discussed in this chapter.

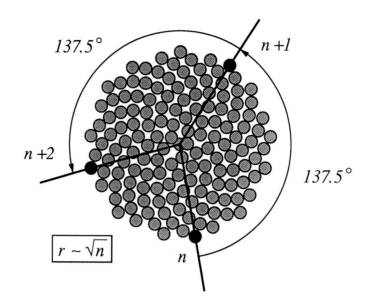

```
#define a 137.5   /* divergence angle */
#include D        /* disk shape specification */

ω  :   A(0)
p₁ :   A(n) :   * → +(a)[f(n∧0.5)~D]A(n+1)
```

Figure 4.1: Pattern of florets in a sunflower head, according to Vogel's formula

4.1   The planar model

Vogel's formula   In order to describe the pattern of florets (or seeds) in a sunflower head, Vogel [154] proposed the formula

$$\phi = n * 137.5°, \qquad r = c\sqrt{n}, \tag{4.1}$$

where:

- n is the ordering number of a floret, counting outward from the center. This is the reverse of floret age in a real plant.

- ϕ is the angle between a reference direction and the position vector of the nth floret in a polar coordinate system originating at the center of the capitulum. It follows that the *divergence angle* between the position vectors of any two successive florets is constant, $\alpha = 137.5°$.

- r is the distance between the center of the capitulum and the center of the nth floret, given a constant scaling parameter c.

a   b   c

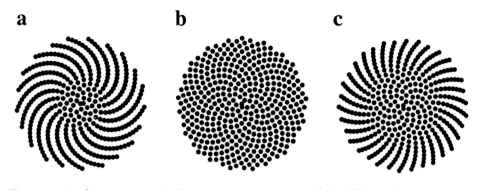

Figure 4.2: Generating phyllotactic patterns on a disk. These three patterns differ only by the value of the divergence angle α, equal to (a) 137.3°, (b) 137.5° (the correct value), and (c) 137.6°.

The distribution of florets described by formula (4.1) is shown in Figure 4.1. The square-root relationship between the distance r and the floret ordering number n has a simple geometric explanation. Assuming that all florets have the same size and are densely packed, the total number of florets that fit inside a disc of radius r is proportional to the disk area. Thus, the ordering number n of the most outwardly positioned floret in the capitulum is proportional to r^2, or $r \sim \sqrt{n}$.

Model justification

The divergence angle of 137.5° is much more difficult to explain. Vogel [154] derives it using two assumptions.

- Each new floret is issued at a fixed angle α with respect to the preceding floret.

- The position vector of each new floret fits into the largest existing gap between the position vectors of the older florets.

Ridley [125] does not object to these basic assumptions, but indicates that they are insufficient to explain the origin of the Fibonacci angle, and points to several arbitrary steps present in Vogel's derivation. He describes the main weakness as follows:

> While it is reasonable to assume that the plant could contain genetic information determining the divergence angle to some extent, it is completely impossible for this alone to fix the divergence angle to the incredible accuracy occurring in nature, since natural variation in biological phenomena is normally rather wide. For example, for the 55- and 89-parastichies to be conspicuous, as occurs in most sunflower heads, d must lie between $\frac{21}{55}$ and $\frac{34}{89}$, a relative accuracy of one part in 1869.

The critical role of the divergence angle α is illustrated in Figure 4.2.

Figure 4.3: Close-up of a daisy capitulum

Parastichies

Although a comprehensive justification of Vogel's formula may require further research, the model correctly describes the arrangement of florets visible in actual capitula. The most prominent feature is two sets of spirals or *parastichies*, one turning clockwise, the other counterclockwise, which are composed of nearest neighboring florets. The number of spirals in each set is always a member of the Fibonacci sequence; 21 and 34 for a small capitulum, up to 89 and 144 or even 144 and 233 for large ones. For example, the capitulum of a daisy (Figure 4.3) exhibits 34 clockwise spirals and 21 counterclockwise spirals, with directions determined by following the spirals outwards from the capitulum center. In the image of a domestic sunflower capitulum (Figure 4.4), one can discern 34 spirals running clockwise and 55 spirals running counterclockwise. The number of perceived spirals depends on the capitulum size expressed in terms of the number of component florets. If the field of attention is limited to a circle approximately 2/3 the size of the entire sunflower capitulum in Figure 4.4, the number of discernible spirals becomes 34 and 21.

Figure 4.4: Domestic sunflower head   ⟶

```
#define S          /* seed shape */
#define R          /* ray floret shape */
#include M N O P  /* petal shapes */

ω  :  A(0)
p₁ :  A(n) :  *  →  +(137.5)[f(n∧0.5)C(n)]A(n+1)
p₂ :  C(n) :  n <= 440                    →  ~S
p₃ :  C(n) :  440 < n & n <= 565          →  ~R
p₄ :  C(n) :  565 < n & n <= 580          →  ~M
p₅ :  C(n) :  580 < n & n <= 595          →  ~N
p₆ :  C(n) :  595 < n & n <= 610          →  ~O
p₇ :  C(n) :  610 < n                      →  ~P
```

The dependence of the number of parastichies on the size of the field of attention is yet another intriguing aspect of spiral phyllotaxis, as pointed out in the following excerpt from a letter by Alan Turing [1] quoted in [18]:

> According to the theory I am working on now there is a continuous advance from one pair of parastichy numbers to another, during the growth of a single plant. ... You will be inclined to ask how one can move continuously from one integer to another. The reason is this — on any specimen there are different ways in which the parastichy numbers can be reckoned; some are more natural than others. During the growth of a plant the various parastichy numbers come into prominence at different stages. One can also observe the phenomenon in space (instead of in time) on a sunflower. It is natural to count the outermost florets as say $21 + 34$, but the inner ones might be counted as $8 + 13$. ... I don't know any really satisfactory account, though I hope to get one myself in about a year's time.

Sunflower head

A complete model of a flower head, suitable for realistic image synthesis, should contain several organs of various shapes. This is easily achieved by associating different surfaces with specific ranges of the index n. For example, consider the L-system that generates the sunflower head (Figure 4.4). The layout of components is specified by production p_1, similar to that of the L-system in Figure 4.1. Productions p_2 to p_7 determine colors and shapes of components as a function of the derivation step number. The entire structure shown in Figure 4.4 was generated in 630 steps. Alternatively, random selection of similar surfaces could have been employed to prevent the excessive regularity of the resulting image.

Other extensions to the basic model consist of varying organ orientation in space and changing their altitude from the plane of the head as a function of n. For example, the sunflower plants included in Figure 4.5 have flowers in four developmental stages: buds, young flowers starting to open, open flowers and older flowers where the petals begin to droop. All flowers are generated using approximately the same number of florets. The central florets are represented by the same surface at each stage. The shape and orientation of surfaces representing petals vary from one stage to another. The plants have been modeled as dibotryoids, with a single signal inducing a basipetal flowering sequence, as described in the previous chapter.

---

[1] To computer scientists, Alan Turing is best known as the inventor of the *Turing machine* [146], which plays an essential role in defining the notion of an algorithm. However, biologists associate Turing's name primarily with his 1952 paper, "The chemical basis of morphogenesis" [147], which pioneered the use of mathematical models in the study of pattern formation and advocated the application of computers to simulate biological phenomena.

Figure 4.5: Sunflower field

Figure 4.6: Zinnias

Other examples   The zinnias (Figures 4.6 and 4.7) illustrate the effect of changing a petal's altitude, size and orientation as a function of n. The height at which a petal is placed decreases by a small amount as n increases. The size of each successive petal is incremented linearly. The orientation is also adjusted linearly by a small angle increment. Thus, petals with small values of index n are placed more vertically, while petals with larger indices n are more horizontal. Although the family Compositae offers the most examples of phyllotactic patterns, the same model can be applied to synthesize images of other flowers, such as water-lilies (Figures 4.8 and 4.9) and roses (Figure 4.10).

Figure 4.7: Close-up of zinnias $\longrightarrow$

Figure 4.8: Water-lily

Figure 4.9: Lily pond

Figure 4.10: Roses

4.2 The cylindrical model

The spiral patterns evident in elongated organs such as pine cones, *Basic model* fir cones and pineapples, can be described by models that position components, in this case scales, on the surface of a cylinder. Van Iterson [75] divides phyllotactic patterns on cylinders into *simple* and *conjugate* ones. In the case of a simple arrangement, all components lie on a single *generative helix*. In contrast, conjugate patterns consist of two or more interleaved helices. This paper discusses simple phyllotactic patterns only. They are generally characterized by the formula

$$\phi = n * \alpha, \qquad r = const, \qquad H = h * n, \qquad (4.2)$$

where:

- n is the ordering number of a scale, counting from the bottom of the cylinder;

- ϕ, r and H are the cylindrical coordinates of the nth scale;

- α is the divergence angle between two consecutive scales (as in the planar case, it is assumed to be constant); and

- h is the vertical distance between two consecutive scales (measured along the main axis of the cylinder).

Implementation
using
L-systems

A parametric L-system that generates the pattern described by formula (4.2) is given in Figure 4.11. The operation of this L-system simulates the natural process of subapical development characterized by sequential production of consecutive modules by the top part of the growing plant or organ. The apex A produces internodes $f(h)$ along the main axis of the modeled structure. Associated with each internode is a disk $\sim D$ placed at a distance r from the axis. This offset is achieved by moving the disk away from the axis using the module $f(r)$, positioned at a right angle with respect to the axis by $\&(90)$. The spiral disk distribution is due to the module $/(a)$, which rotates the apex around its own axis by the divergence angle in each derivation step.

Analysis of
model geometry

In the planar model, the constant divergence angle $\alpha = 137.5°$ is found across a large variety of flower heads. The number of perceived parastichies is determined by the capitulum size, and it changes as the distance from the capitulum center increases. In contrast, a phyllotactic pattern on the surface of a cylinder is uniform along the entire cylinder length. The number of evident parastichies depends on the values of parameters α and h. The key problem, both from the viewpoint of understanding the geometry of the pattern and applying it to generate synthetic images, is to express the divergence angle α and the vertical displacement h as a function of the numbers of evident parastichies encircling the cylinder in the clockwise and counterclockwise directions. A solution to this problem was proposed by van Iterson [75] and reviewed by Erickson [36]. Our presentation closely follows that of Erickson.

The phyllotactic pattern can be explained in terms of circles packed on the surface of the cylinder. An evident parastichy consists of a sequence of tangent circles, the ordering numbers of which form an arithmetic sequence with difference m. The number m is referred to as the *parastichy order*. Thus, the circles on the cylinder surface may be arranged in two congruent 2-parastichies, five congruent 5-parastichies, and so on. The angular displacement between two consecutive circles in an m-parastichy is denoted by δ_m. By definition, δ_m belongs to the range $(-\pi, \pi]$ radians. The relation between the angular displacement δ_m and the divergence angle α is expressed by the equation

$$\delta_m = m\alpha - \Delta_m 2\pi, \qquad (4.3)$$

where Δ_m is an integer called the *encyclic number*. It is the number of turns around the cylinder, rounded to the nearest integer, which the generative helix describes between two consecutive points of the m-parastichy.

Usually, one can perceive two series of parastichies running in opposite directions (Figure 4.11). The second parastichy satisfies an equation analogous to (4.3):

$$\delta_n = n\alpha - \Delta_n 2\pi \qquad (4.4)$$

Consider the m- and n-parastichies starting at circle 0. In their paths across the cylinder, they will intersect again at circle mn. Assume

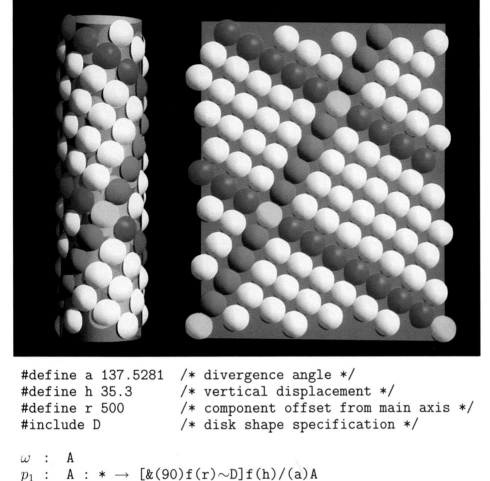

```
#define a 137.5281    /* divergence angle */
#define h 35.3        /* vertical displacement */
#define r 500         /* component offset from main axis */
#include D            /* disk shape specification */


ω  :  A
p₁ :  A : * → [&(90)f(r)~D]f(h)/(a)A
```

Figure 4.11: Parastichies on the surface of a cylinder and on the unrolled cylinder. The L-system generates the cylindrical pattern.

that m and n are relatively prime; otherwise the phyllotactic pattern would have to contain several circles lying at the same height H and, contrary to the initial assumption, would not be simple. The circle mn is the first point of intersection between the m-parastichy and the n-parastichy above circle 0. Consequently, the path from circle 0 to mn along the m-parastichy, and back to 0 along the n-parastichy, encircles the cylinder exactly once. The section of m-parastichy between circles 0 and mn consists of $n+1$ circles (including the endpoints), so the angular distance between the circles 0 and mn is equal to $n\delta_m$. Similarly, the distance between circles 0 and mn, measured along the n-parastichy, can be expressed as $m\delta_n$. As a result,

$$n\delta_m - m\delta_n = \pm 2\pi. \tag{4.5}$$

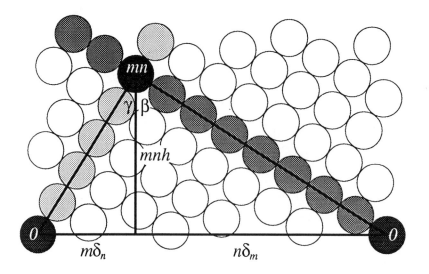

Figure 4.12: An opposite parastichy triangle (as in Erickson [36, Fig. 3.8]). The base is formed by the circumference of the cylinder. The sides are formed by the parastichies.

The signs in equation (4.5) correspond to the assumption that the spirals encircle the cylinder in opposite directions; thus one of the values δ is positive and the other one is negative. Substituting the right sides of equations (4.3) and (4.4) for δ_m and δ_n yields

$$n\Delta_m - m\Delta_n = \pm 1. \tag{4.6}$$

To further analyze the pertinent geometric relationships, the cylinder is cut along the vertical line passing through the center of circle 0 and "unrolled" (Figure 4.11). The two parastichies and the circumference of the cylinder passing through point 0 form a triangle as shown in Figure 4.12. The perpendicular to the base from point mn divides this triangle into two right triangles. If d denotes the diameter of the circles, then

$$(n\delta_m)^2 + (mnh)^2 = (nd)^2$$

and

$$(m\delta_n)^2 + (mnh)^2 = (md)^2.$$

The above system of equations can be solved with respect to h and d:

$$h = \sqrt{(\delta_m^2 - \delta_n^2)/(n^2 - m^2)} \tag{4.7}$$

$$d = \sqrt{(n^2\delta_m^2 - m^2\delta_n^2)/(n^2 - m^2)} \tag{4.8}$$

or, taking into consideration equation (4.5),

$$d = \sqrt{2\pi(n\delta_m + m\delta_n)/(n^2 - m^2)}. \tag{4.9}$$

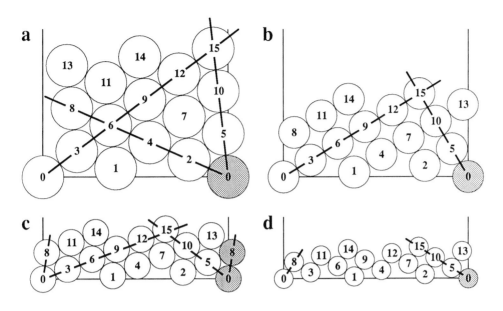

Figure 4.13: Patterns of tangent circles drawn on the surface of a cylinder as a function of circle diameter

The problem is to determine values of δ_m and δ_n. They are not simply functions of parameters m and n. Figure 4.13 shows that, for a given m and n, the values of δ_m and δ_n can be chosen from a certain range, yielding parastichies of different steepness. In order to determine this range, observe that at its limits the phyllotactic pattern changes; one previously evident parastichy disappears and another is formed. Thus, at the range limit, *three* evident parastichies coexist. It follows from Figure 4.13 that at one end of the range the third parastichy has order $|m-n|$, and at the other end it has order $(m+n)$. Three coexisting parastichies imply that each circle is tangent to six other circles. In other words, all circles lie in the vertices of a regular hexagonal grid, as seen in Figure 4.13a and c. Consequently, the angle $\beta + \gamma$ at vertex mn (Figure 4.12) is equal to $2\pi/3$. Expressing the base of the triangle in terms of its two sides and their included angle results in

$$(2\pi)^2 = (nd)^2 + (md)^2 - 2(nd)(md)\cos(2\pi/3)$$

or, after simplification,

$$d = 2\pi/\sqrt{m^2 + mn + n^2}. \qquad (4.10)$$

Equations (4.9) and (4.10) yield

$$n\delta_m + m\delta_n = 2\pi(n^2 - m^2)/(m^2 + mn + n^2). \qquad (4.11)$$

Solving the system of equations (4.5) and (4.11) with respect to δ_m and δ_n produces

$$\delta_m = \pi(m + 2n)/(m^2 + mn + n^2) \qquad (4.12)$$

and

$$\delta_n = \pi(2m + n)/(m^2 + mn + n^2). \qquad (4.13)$$

Given the values of δ_m and δ_n, the divergence angle α can be found from either equation (4.3) or (4.4), assuming that the encyclic numbers Δ_m or Δ_n are known. It follows from the definition that these numbers are the *smallest* positive integers satisfying equation (4.6). A systematic method for solving this equation, based on the theory of continuous fractions, is presented by van Iterson [75]. Erickson [36] points out that in practice the solution can often be found by guessing. Another possibility is to look for the smallest pair of numbers (Δ_m, Δ_n) satisfying (4.6) using a simple computer program.

Pattern construction

In conclusion, a phyllotactic pattern characterized by a pair of numbers (m, n) can be constructed as follows:

1. Find Δ_m and Δ_n from equation (4.6).

2. Find the range of admissible values of the angular displacements δ_m and δ_n. The limits can be obtained from equations (4.12) and (4.13) using the values of m and n for one limit, and the pair $(\min\{m, n\}, |m - n|)$ for the other.

3. For a chosen pair of admissible displacement values δ_m and δ_n, calculate the divergence angle α from equation (4.3) or (4.4) and the vertical displacement h from equation (4.8).

4. Find the diameter d of the circles from equation (4.8).

The diameter d does not enter directly in any formula used for image synthesis, but serves as an estimate of the size of surfaces to be incorporated in the model. This algorithm was applied to produce Table 4.1 showing parameter values for which three parastichies coexist. Given a pattern with two parastichies, this table provides the limits of the divergence angle α. For example, a (5,8) pattern can be formed for values of α ranging from 135.918365° to 138.139542°, which correspond to the patterns (3,5,8) and (5,8,13), respectively.

Triple-contact patterns

Further information relating the divergence angle α to the vertical displacement h for various phyllotactic patterns is shown in Figure 4.14. The arcs represent parameters of patterns with two parastichies (m, n). The branching points represent parameters of patterns with three parastichies $(m, n, m + n)$.

$m,n,m+n$	α (degrees)	h	d
$(1, 1, 2)$	180.000000	1.81380	-
$(1, 2, 3)$	128.571426	0.777343	2.374821
$(1, 3, 4)$	96.923073	0.418569	1.742642
$(2, 3, 5)$	142.105270	0.286389	1.441462
$(1, 4, 5)$	77.142860	0.259114	1.371104
$(3, 4, 7)$	102.162163	0.147065	1.032949
$(3, 5, 8)$	135.918365	0.111049	0.897598
$(2, 5, 7)$	152.307693	0.139523	1.006115
$(1, 5, 6)$	63.870968	0.175529	1.128493
$(4, 5, 9)$	79.672134	0.089203	0.804479
$(4, 7, 11)$	98.709671	0.058510	0.651536
$(3, 7, 10)$	107.088600	0.068878	0.706914
$(3, 8, 11)$	131.752579	0.056097	0.637961
$(5, 8, 13)$	138.139542	0.042181	0.553204
$(5, 7, 12)$	150.275223	0.049921	0.601820
$(2, 7, 9)$	158.507462	0.081215	0.767613

Table 4.1: Cylinder formula values for triple-contact patterns

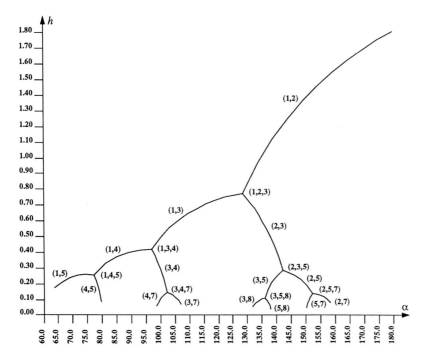

Figure 4.14: The vertical displacement h as a function of the divergence angle α for various phyllotactic patterns

Figure 4.15: Pineapples

Pineapple

Spruce cones

Sedge

Models of fruits synthesized using the cylindrical model are shown in Figures 4.15 and 4.16.   The pineapple (Figure 4.15) is an example of a pattern where a given scale has six neighbors, which belong to 5-, 8- and 13-parastichies. The corresponding divergence angle α is equal to 138.139542°. The spruce cones (Figure 4.16) were generated using the values $m = 5$, $n = 8$ and $\alpha = 137.5°$ (the divergence angle α for a $(5, 8)$-parastichy pattern belongs to the interval 135.918365° to 138.139542°). From these values, h and d were calculated as a function of the radius of the cylinder. The effect of closing the bottom and top of the pineapple and spruce cones was achieved by decreasing the diameter of the cylinder and the size of the scales.

A variant of the model of phyllotaxis on a cylinder can be used to model organs that are conical rather than cylindrical in shape. For example, Figure 4.17 shows a model of the sedge *Carex laevigata*. L-system 4.1 generates the male spike. Production p_1 specifies the basic

Figure 4.16: Spruce cones

Figure 4.17: *Carex laevigata*: the male spike, the entire shoot, the female spike

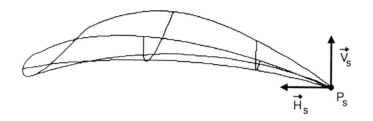

Figure 5.1: Specification of an appendage

of the turtle, and is rotated to align its heading and up vectors with the corresponding vectors of the turtle. If a surface represents an internal part of the structure such as an internode, a distinction between the entry and exit contact points is made.

Examples   The majority of organs presented in this book have been modeled this way. The petals of sunflowers, zinnias, water-lilies and roses shown in Chapter 4 provide good examples. Figure 5.2 illustrates an additional improvement in the appearance of organs, made possible by the application of textures to the surfaces of leaves, flowers and vine branches.

5.2   Developmental surface models

Predefined surfaces do not "grow." String symbols can be applied to control such features as the overall color and size of a surface, but the underlying shape remains the same. In order to simulate plant development fully, it is necessary to provide a mechanism for changing the shape as well as the size of surfaces in time. One approach is to *Contour tracing* trace surface boundaries using the turtle and fill the resulting polygons. A sample L-system is given below:

$$
\begin{aligned}
\omega &: \quad L \\
p_1 &: \quad L \rightarrow \{ -FX + X - FX - | - FX + X + FX \} \\
p_2 &: \quad X \rightarrow FX
\end{aligned}
$$

Production p_1 defines leaf L as a closed planar polygon. The braces { and } indicate that this polygon should be filled. Production p_2 *Fern* increases the lengths of its edges linearly. The model of a fern shown in Figure 5.3 incorporates leaves generated using this method, with the angle increment equal to 20°. Note the phase effect due to the "growth" of polygons in time. A similar approach was taken to generate the leaves, flowers and fruits of *Capsella bursa-pastoris* (Figure 3.5 on page 74).

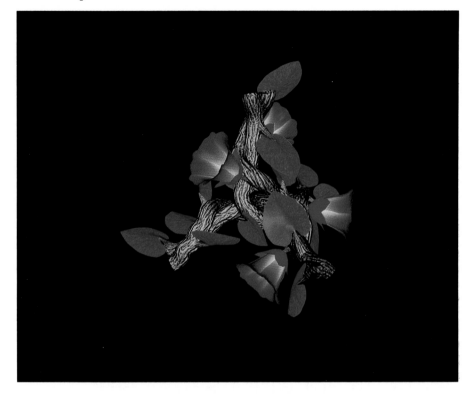

Figure 5.2: *Maraldi figure* by Greene [54]

Figure 5.3: The fern

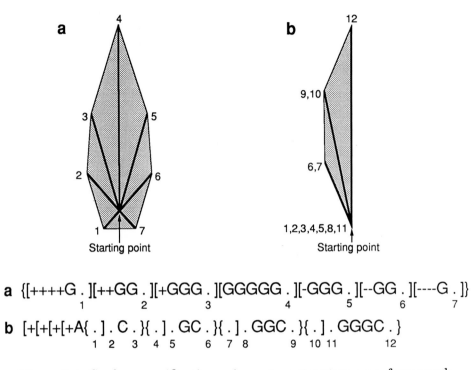

a {[++++G .][++GG .][+GGG .][GGGGG .][-GGG .][--GG .][----G .]}
            1   2   3   4   5   6   7

b [+[+[+[+A{ .] . C . }{ .] . GC . }{ .] . GGC . }{ .] . GGGC . }
            1 2   3   4 5   6 7 8   9 10 11   12

Figure 5.4: Surface specification using a tree structure as a framework

Framework approach

In practice, the tracing of polygon boundaries produces acceptable effects only in the case of small, flat surfaces. In other cases it is more convenient to use a tree structure as a *framework*. Polygon vertices are specified by a sequence of turtle positions marked by the dot symbol (.). An example is given in Figure 5.4a. The letter G has been used instead of F to indicate that the segments enclosed between the braces should not be interpreted as the edges of the constructed polygon. The numbers correspond to the order of vertex specification by the turtle.

Cordate leaf

Figure 5.5 shows the development of a *cordate leaf* modeled using this approach. The axiom contains symbols A and B, which initiate the left-hand and right-hand sides of the blade. Each of the productions p_1 and p_2 creates a sequence of axes starting at the leaf base and gradually diverging from the midrib. Production p_3 increases the lengths of the axes. The axes close to the midrib are the longest since they were created first. Thus, the shape of this leaf is yet another manifestation of the phase effect. The leaf blade is defined as a union of triangles rather than a single polygon. Such triangulation is advantageous if the blade bends, for example due to tropism (Chapter 2). Figure 5.4b provides an additional illustration of the model by magnifying the left side of the leaf after four derivation steps.

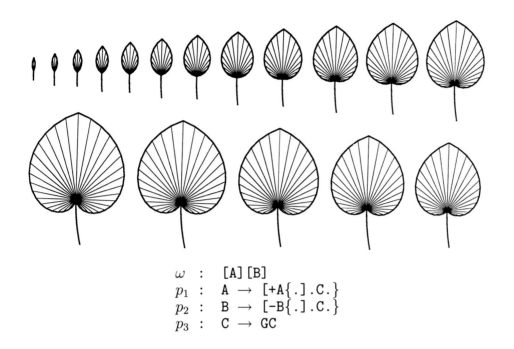

$$\omega \quad : \quad \texttt{[A][B]}$$
$$p_1 \quad : \quad \texttt{A} \; \rightarrow \; \texttt{[+A\{.].C.\}}$$
$$p_2 \quad : \quad \texttt{B} \; \rightarrow \; \texttt{[-B\{.].C.\}}$$
$$p_3 \quad : \quad \texttt{C} \; \rightarrow \; \texttt{GC}$$

Figure 5.5: Developmental sequence of a cordate leaf generated using an L-system

The described method makes it possible to define a variety of leaves. *Simple leaves*
Their shapes depend strongly on the growth rates of segments. For example, a family of simple leaves and the corresponding parametric L-system are shown in Figure 5.6.

According to production p_1, in each derivation step apex $A(t)$ extends the main leaf axis by segment $G(LA, RA)$ and creates a pair of lateral apices $B(t)$. New lateral segments are added by production p_2. Parameter t, assigned to apices B by production p_1, plays the role of "growth potential" of the branches. It is decremented in each derivation step by a constant PD, and stops production of new lateral segments upon reaching 0. Segment elongation is captured by production p_3.

For the purpose of analysis, it is convenient to divide a leaf blade into two areas. In the basal area, the number of lateral segments is determined by the initial value of growth potential t and constant PD. Since the initial value of t assigned to apices B increases towards the leaf apex, the consecutive branches contain more and more segments. On the other hand, branches in the apical area exist for too short a time to reach their limit length. Thus, while traversing the leaf from the base towards the apex, the actual number of segments in a branch first increases, then decreases. As a result of these opposite tendencies, the leaf reaches its maximum width near the central part of the blade.

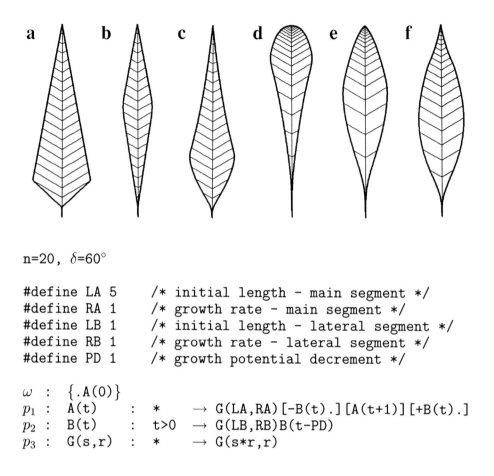

n=20, $\delta=60°$

```
#define LA 5      /* initial length - main segment */
#define RA 1      /* growth rate - main segment */
#define LB 1      /* initial length - lateral segment */
#define RB 1      /* growth rate - lateral segment */
#define PD 1      /* growth potential decrement */
```

ω : $\{.A(0)\}$
p_1 : A(t)   : * $\rightarrow$ G(LA,RA)[-B(t).][A(t+1)][+B(t).]
p_2 : B(t)   : t>0 $\rightarrow$ G(LB,RB)B(t-PD)
p_3 : G(s,r) : * $\rightarrow$ G(s*r,r)

Figure 5.6: A family of simple leaves generated using a parametric L-system

Figure	LA	RA	LB	RB	PD
a	5	1.0	1.0	1.00	0.00
b	5	1.0	1.0	1.00	1.00
c	5	1.0	0.6	1.06	0.25
d	5	1.2	10.0	1.00	0.50
e	5	1.2	4.0	1.10	0.25
f	5	1.1	1.0	1.20	1.00

Table 5.1: Values of constants used to generate simple leaves

Figure 5.7: A rose in a vase

Table 5.1 lists the values of constants corresponding to particular shapes. *Shape control*
For a given derivation length, the exact position of the branch with the
largest number of segments is determined by PD. If PD is equal to 0,
all lateral branches have an unlimited growth potential, and the basal
part of the leaf does not exist (Figure 5.6a). If PD equals 1, the basal
and apical parts contain equal numbers of lateral branches (Figures 5.6
b and f). Finer details of the leaf shape are determined by the growth
rates. If the main axis segments and the lateral segments have the same
growth rates ($RA = RB$), the edges of the apical part of the leaf are
straight (Figures 5.6 a and b). If RA is less than RB, the segments
along the main axis elongate at a slower rate than the lateral segments,
resulting in a concave shape of the apical part (Figures 5.6 c and f). In
the opposite case, with RA greater than RB, the apical part is convex
(Figures 5.6 d and e). The curvature of the basal edges can be analyzed
in a similar way.

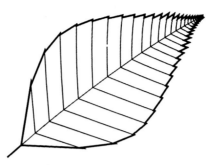

n=25, δ=60°

```
#define LA 5      /* initial length - main segment */
#define RA 1.15   /* growth rate - main segment */
#define LB 1.3    /* initial length - lateral segment */
#define RB 1.25   /* growth rate - lateral segment */
#define LC 3      /* initial length - marginal notch */
#define RC 1.19   /* growth rate - marginal notch */
```

ω : [{A(0,0).}][{A(0,1).}]
p_1 : A(t,d)   : d=0 → .G(LA,RA).[+B(t)G(LC,RC,t).}]
                                      [+B(t){.]A(t+1,d)
p_2 : A(t,d)   : d=1 → .G(LA,RA).[-B(t)G(LC,RC,t).}]
                                      [-B(t){.]A(t+1,d)
p_3 : B(t)   : t>0 → G(LB,RB)B(t-1)
p_4 : G(s,r)   : *   → G(s*r,r)
p_5 : G(s,r,t) : t>1 → G(s*r,r,t-1)

Figure 5.8: A rose leaf

Rose leaf

Figure 5.7 shows a long-stemmed rose with the leaves modeled according to Figure 5.8. The L-system combines the concepts explored in Figures 5.5 and 5.6. The axiom contains modules $A(0,0)$ and $A(0,1)$, which initiate the left-hand and right-hand side of the leaf. The development of the left side will be examined in detail. According to production p_1, in each derivation step apex $A(t,0)$ extends the midrib by internodes $G(LA,RA)$ and creates two colinear apices $B(t)$ pointing to the left. Further extension of the lateral axes is specified by production p_3. The leaf blade is constructed as a sequence of trapezoids, with two vertices lying on the midrib and the other two vertices placed at the endpoints of a pair of lateral axes formed in consecutive derivation steps. The module $G(LC,RC,t)$ introduces an offset responsible for the formation of notches at the leaf margin. Production p_4 describes the elongation of internodes responsible for overall leaf shape, while production p_5 controls the size of the notches. The development of the right side of the blade proceeds in a similar manner, with production p_2
```
```

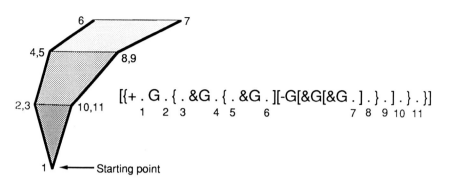

Figure 5.9: Surface specification using stacked polygons

recreating the midrib and creating lateral apices pointing to the right. The bending of the midrib to the right is a result of tropism.

Nested polygons

In the examples discussed so far, the turtle specifies the vertices of one polygon, then moves on to the next. Further flexibility in surface definition can be achieved by interleaving vertex specifications for different polygons. The turtle interpretation of the braces is redefined in the following way. A string containing nested braces is evaluated using two data structures, an *array* of vertices representing the current polygon and a polygon *stack*. At the beginning of string interpretation, both structures are empty. The symbols {, } and . are then interpreted as follows:

{   Start a new polygon by pushing the current polygon on the polygon stack and creating an empty current polygon.

.   Append the new vertex to the current polygon.

}   Draw the current polygon using the specified vertices, then pop a polygon from the stack and make it the current polygon.

An example of string interpretation involving nested braces is given in Figure 5.9.

Lily-of-the-valley

The above technique was applied to construct the flowers of the lily-of-the-valley shown in Figure 3.4 (page 72), and magnified in Figure 5.10. A flower is represented by a polygon mesh consisting of five sequences of trapezoids spread between pairs of curved lines that emanate radially from the flower base. A single sequence is generated by the following L-system:

$$\omega \; : \; [X(36)A]/(72)[X(36)B]$$
$$p_1 : \; A \quad : \; * \; \rightarrow \; [\&GA\{.].$$
$$p_2 : \; B \quad : \; * \; \rightarrow \; B\&.G.\}$$
$$p_3 : \; X(a) : \; * \; \rightarrow \; X(a+4.5)$$

Chapter 6

Animation of plant development

The sequences of images used in Chapters 3 and 5 to illustrate the development of inflorescences and compound leaves suggest the possibility of using computer animation to visualize plant development. From a practical perspective, computer animation offers several advantages over traditional time-lapse photography.

Motivation

- Photography is sensitive to imperfections in the underlying experiment. A disease or even a temporary wilting of a plant may spoil months of filming.

- In nature, developmental processes are often masked by other phenomena. For example, the growth of leaves can be difficult to capture because of large changes in leaf positions during the day. Similarly, positions of tree branches may be affected by wind. Computer animation makes it possible to abstract from these distracting effects.

- Animation can be used when time-lapse photography is impractical because of the long development time of plants (e.g. years in the case of trees).

- In time-lapse photography, the initial position of the camera and the light conditions must be established a priori, without knowing the final shape of the plant. In computer animation all developmental stages of the modeled plant are known in advance, allowing for optimal selection of the model view.

- Animation can be applied to visualize developmental mechanisms that cannot be observed directly in real plants, such as the propagation of hormones and nutrients.

- Animation offers an unprecedented means for visualizing the development of extinct plants on the basis of paleobotanical data.

Discrete character of L-systems

The original formalism of L-systems provides a model of development that is discrete both in time and space. Discretization in time implies that the model states are known only at specific time intervals. Discretization in space means that the range of model states is limited to a finite number. Parametric L-systems remove the latter effect by assigning continuous attributes to model components. However, the model states are still known only in discrete time intervals. This presents a problem in animation, where a smooth progression of the developing forms is desirable.

This last statement requires further clarification. The very nature of animation is to produce the impression of continuous motion by displaying a sequence of still frames, captured at fixed time intervals. Why is a continuous model of plant development needed if it will be used to generate a fixed sequence of images in the final account? Wouldn't it be enough to retain the standard definition of L-systems and assume time slices fine enough to produce the desired progression of forms? This approach, while feasible and useful, has three major drawbacks.

- A model can be constructed assuming longer or shorter time intervals, but once the choice has been made, the time step is a part of the model and cannot be changed easily. From the viewpoint of computer animation it is preferable to control the time step by a single parameter, decoupled from the underlying L-system.

- The continuity criteria responsible for the smooth progression of shapes during animation can be specified more easily in the continuous time domain.

- It would be conceptually elegant to separate model development, defined in continuous time, from model observation, taking place in discrete intervals.

A developmental process is viewed as consisting of periods of continuous module expansion delimited by instantaneous module divisions. Special conditions are imposed to preserve the shape and growth rates of the organism during these qualitative changes. An analogy can be drawn to the theory of morphogenesis advanced by Thom [142], who viewed shape formation as a piecewise continuous process with singularities called *catastrophes*.

Formally, development taking place in continuous time is captured using the formalism of *timed DOL-systems*. The key difference between these L-systems and the types of L-systems considered so far lies in the definition of the derivation function. In "ordinary" L-systems, the derivation length is expressed as the number of derivation steps. In timed DOL-systems, the derivation length is associated with the total elapsed time since the beginning of the observation.

6.1 Timed DOL-systems

Let V be an alphabet and R the set of positive real numbers (including 0). The pair $(a, \tau) \in V \times R$ is referred to as the *timed letter a*, and the number τ is called the *age* of a. A sequence of timed letters, $x = (a_1, \tau_1) \ldots (a_n, \tau_n) \in (V \times R)^*$, is called a *timed word* over alphabet V.

A *timed DOL-system (tDOL-system)* is a triplet $G = \langle V, \omega, P \rangle$,   *Definition*
where

- V is the alphabet of the L-system,

- $\omega \in (V \times R)^+$ is a nonempty timed word over V, called the initial word, and

- $P \subset (V \times R) \times (V \times R)^*$ is a finite set of productions.

Instead of writing $((a, \beta), (b_1, \alpha_1) \ldots (b_n, \alpha_n)) \in P$, the notation $(a, \beta) \rightarrow (b_1, \alpha_1) \ldots (b_n, \alpha_n)$ is used. The parameter β is referred to as the *terminal age* of the letter a, and each parameter α_i is the *initial age* assigned to the letter b_i by production P. The following assumptions are made:

C1. For each letter $a \in V$ there exists exactly one value $\beta \in R$ such that (a, β) is the predecessor of a production in P.

C2. If (a, β) is a production predecessor and (a, α) is a timed letter that occurs in the successor of some production in P, then $\beta > \alpha$.

According to these conditions, each letter has a uniquely defined terminal age. Furthermore, an initial age assigned to a letter by a production must be smaller than its terminal age, *i.e.*, its *lifetime* $(\beta - \alpha)$ must be positive.

Let $(a, \beta) \rightarrow (b_1, \alpha_1) \ldots (b_n, \alpha_n)$ be a production in P. A function   *Derivation*
$\mathcal{D} : ((V \times R)^+ \times R) \rightarrow (V \times R)^*$ is called a *derivation function* if it has the following properties:

A1. $\mathcal{D}(((a_1, \tau_1) \ldots (a_n, \tau_n)), t) = \mathcal{D}((a_1, \tau_1), t) \ldots \mathcal{D}((a_n, \tau_n), t)$

A2. $\mathcal{D}((a, \tau), t) = (a, \tau + t)$, if $\tau + t \leq \beta$

A3. $\mathcal{D}((a, \tau), t) = \mathcal{D}((b_1, \alpha_1) \ldots (b_n, \alpha_n), t - (\beta - \tau))$, if $\tau + t > \beta$

A derivation in a timed DOL-system is defined in terms of two types of time variables. Global time t is common to the entire word under consideration, while local age values τ_i are specific to each letter a_i.

Axiom **A1** identifies t as the variable that synchronizes the entire development, and specifies that the lineages of all letters can be considered independently from each other (thus, no interaction between letters is assumed). With the progress of time t, each letter "grows older" until its terminal age is reached (axiom **A2**). At this moment subdivision occurs and new letters are produced with initial age values specified by the corresponding production (axiom **A3**). Condition **C1** guarantees that the subdivision time is defined unambiguously, hence the development proceeds in a deterministic fashion. Condition **C2** guarantees that, for any value of time t, the recursive references specified by axiom **A3** will eventually end.

Anabaena The above concepts are examined by formulating a timed DOL-system that simulates the development of a vegetative part of the *Anabaena catenula* filament. Given the discrete model expressed by equation (1.1) on page 5, the corresponding tDOL-system is as follows:

$$
\begin{aligned}
\omega : \quad & (a_r, 0) \\
p_1 : \quad & (a_r, 1) &\rightarrow\quad& (a_l, 0)(b_r, 0) \\
p_2 : \quad & (a_l, 1) &\rightarrow\quad& (b_l, 0)(a_r, 0) \\
p_3 : \quad & (b_r, 1) &\rightarrow\quad& (a_r, 0) \\
p_4 : \quad & (b_l, 1) &\rightarrow\quad& (a_l, 0)
\end{aligned}
\qquad (6.1)
$$

In accordance with the discrete model, it is assumed that all cells have the same lifetime, equal to one time unit. The derivation tree is shown in Figure 6.1. The nodes of the tree indicate production applications specified by axiom **A3**, and the triangular "arcs" represent the continuous aging processes described by axiom **A2**. The vertical scale indicates global time. For example, at time $t = 2.75$ the filament consists of three cells, b_l, a_r and a_r, whose current age is equal to 0.75.

According to the definition of time intervals corresponding to axioms **A2** and **A3**, a production is applied *after* the age $\tau + t$ exceeds the terminal age. Consequently, at division time the "old" cells still exist. For example, at time $t = 2.0$ the filament consists of two cells, a_l and b_r, both of age $\tau = 1$.

The above L-system can be simplified by considering cells of type b as young forms of the cells of type a. This is suggested by Figure 6.1 where cells b simply precede cells a in time. The simplified L-system has two productions:

$$
\begin{aligned}
p_1 : \quad & (a_r, 2) &\rightarrow\quad& (a_l, 1)(a_r, 0) \\
p_2 : \quad & (a_l, 2) &\rightarrow\quad& (a_l, 0)(a_r, 1)
\end{aligned}
\qquad (6.2)
$$

The corresponding derivation tree starting from cell $(a_r, 1)$ is shown in Figure 6.2. Note the similarity to the tree from the previous example.

Model
observation Whether a natural developmental process or its mathematical model is considered, the choice of observation times and the act of observation should not affect the process itself. In other terms, each derived word should depend only on the *total elapsed time t*, and not on the partition

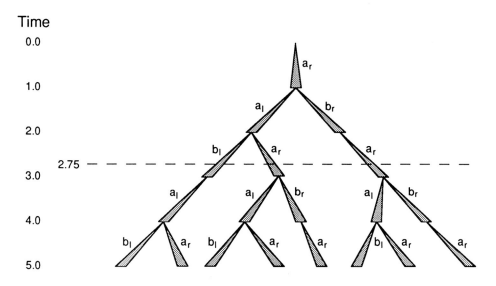

Figure 6.1: Derivation tree representing the continuous development of *Anabaena catenula* described by the L-system is equation (6.1). Sections of the triangles represent cell ages.

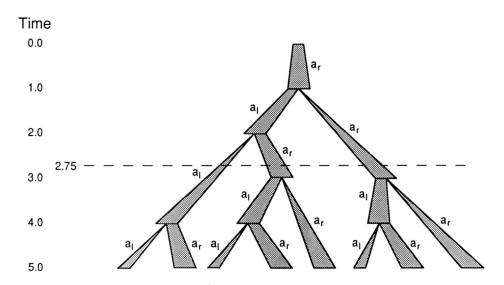

Figure 6.2: Derivation tree representing the continuous development of *Anabaena catenula* corresponding to the rules specified in equation (6.2)

of t into intervals. The following theorem shows that timed DOL-systems satisfy this postulate.

Theorem. Let $G = \langle V, \omega, P \rangle$ be a tDOL-system, and $x \in (V \times R)^+$ be a timed word over V. For any values of $t_a, t_b \geq 0$, the following holds:

$$\mathcal{D}(\mathcal{D}(x, t_a), t_b) = \mathcal{D}(x, t_a + t_b)$$

Proof. Let us first consider the special case where the word x consists of a single timed letter, (a_0, τ_0), and all productions in the set P take single letters into single letters. According to condition **C1**, there exists a unique sequence of productions from P such that:

$$(a_i, \beta_i) \rightarrow (a_{i+1}, \alpha_{i+1}), \quad i = 0, 1, 2, \ldots$$

Let (a_k, τ_k) be the result of the derivation of duration t_a that starts from (a_0, τ_0). According to axioms **A2** and **A3**, and assuming that $t_a > \beta_0 - \tau_0$, this derivation can be represented in the form

$$
\begin{aligned}
\mathcal{D} \ & ((a_0, \tau_0), t_a) \\
= \ & \mathcal{D}((a_1, \alpha_1), \ t_a - (\beta_0 - \tau_0)) \\
= \ & \mathcal{D}((a_2, \alpha_2), \ t_a - (\beta_0 - \tau_0) - (\beta_1 - \alpha_1)) \\
= \ & \cdots \\
= \ & \mathcal{D}((a_k, \alpha_k), \ t_a - (\beta_0 - \tau_0) - (\beta_1 - \alpha_1) - \ldots - (\beta_{k-1} - \alpha_{k-1})) \\
= \ & (a_k, \tau_k),
\end{aligned}
$$

where

$$\tau_k = \alpha_k + [t_a - (\beta_0 - \tau_0) - \sum_{i=1}^{k-1} (\beta_i - \alpha_i)].$$

Since the sequence of recursive calls can be terminated only by an application of axiom **A2**, the index k and the age τ_k satisfy the inequality

$$\alpha_k < \tau_k \leq \beta_k.$$

Due to condition **C2**, such an index k always exists and is unique.

Let us now consider a derivation of duration $t_b > \beta_k - \tau_k$ that starts from (a_k, τ_k). Following the same reasoning, the result can be represented as (a_m, τ_m), where

$$\tau_m = \alpha_m + [t_b - (\beta_k - \tau_k) - \sum_{i=k+1}^{m-1} (\beta_i - \alpha_i)]$$

and

$$\alpha_m < \tau_m \leq \beta_m.$$

By substituting the value of τ_k into the formula for τ_m, we obtain after simple transformations

$$\tau_m = \alpha_m + [(t_a + t_b) - (\beta_0 - \tau_0) - \sum_{i=1}^{m-1} (\beta_i - \alpha_i)].$$

Thus, the timed letter (a_m, τ_m') also results from the derivation of duration $t_a + t_b$ starting with (a_0, τ_0):

$$(a_m, \tau_m) = \mathcal{D}(\mathcal{D}((a_0, \tau_0), t_a), t_b) = \mathcal{D}((a_0, \tau_0)x, t_a + t_b).$$

So far, we have considered only the case

$$t_a > \beta_0 - \tau_0, \qquad t_b > \beta_k - \tau_k.$$

Two other cases, namely,

$$0 \le t_a \le \beta_0 - \tau_0, \qquad t_b > \beta_k - \tau_k$$

and

$$t_a > \beta_0 - \tau_0, \qquad 0 \le t_b \le \beta_k - \tau_k$$

can be considered in a similar way. The remaining case,

$$0 \le t_a \le \beta_0 - \tau_0, \qquad 0 \le t_b \le \beta_k - \tau_k,$$

is a straightforward consequence of condition **C2**. This completes the proof of the special case. In general, a derivation that starts from a word $(a_1, \tau_1) \ldots (a_n, \tau_n)$ can be considered as n separate derivations, each starting from a single letter. This observation applies not only to the initial word specified at time $t = 0$, but also to any intermediate word generated during the derivation. Consequently, any *path* in the derivation tree can be considered as a sequence of mappings that takes single letters into single letters. Application of the previous reasoning separately to every path concludes the proof. $\square$

6.2   Selection of growth functions

Timed L-systems capture qualitative changes in model topology corresponding to cell (or, in general, module) divisions, and return the age of each module as a function of the global time t. In order to complete model definition, it is also necessary to specify the shape of each module as a function of its age. Potentially, such growth functions can be estimated experimentally by observing real organisms [72, 73]. However, if detailed data is not available, growth functions can be selected from an appropriate class by choosing parameters so that the animation is smooth. This approach can be viewed as more than an *ad hoc* technique for constructing acceptable animated sequences. In fact, Thom presents it as a general methodology for studying morphogenesis [142, page 4]:

> The essence of the method to be described here consists in supposing a priori the existence of a differential model underlying the process to be studied and, without knowing explicitly what the model is, deducing from the single assumption of its existence conclusions relating to the nature of the singularities of the process.

A technique for computing parameters of growth functions in the case of nonbranching filaments and simple branching structures is given below.

6.2.1   Development of nonbranching filaments

In a simple case of geometric interpretation of timed L-systems, symbols represent cells that elongate during their lifetime and divide upon reaching terminal age. Model geometry does not change suddenly, which means that

- the length of each cell is a continuous function of time, and

- the length of a cell before subdivision is equal to the sum of the lengths of the daughter cells.

Continuity requirements
The above *continuity requirements* are formalized in the context of a tDOL-system $G = \langle V, \omega, P \rangle$ as follows:

R1. Let $[\alpha_{min}, \beta]$ describe the life span of a timed letter $a \in V$. The age α_{min} is the minimum of the initial age values assigned to a by the axiom ω or by some production in P. The terminal age β is determined by the predecessor of the production acting on symbol a. The *growth function* $g(a, \tau)$, which specifies the length of cell a as a function of age τ, must be a continuous function of parameter τ in the domain $[\alpha_{min}, \beta]$.

R2. For each production $(a, \beta) \rightarrow (b_1, \alpha_1) \ldots (b_n, \alpha_n)$ in P the following equality holds:

$$g(a, \beta) = \sum_{i=1}^{n} g(b_i, \alpha_i) \qquad (6.3)$$

In practice, requirement **R1** is used to select the class of growth functions under consideration, and the equations resulting from requirement **R2** are used to determine the parameters in function definitions.

Linear growth
For example, in the case of the timed L-system specified in equation (6.2), requirement **R2** takes the form

$$\begin{aligned} g(a_r, 2) &= g(a_l, 1) + g(a_r, 0) \\ g(a_l, 2) &= g(a_l, 0) + g(a_r, 1). \end{aligned} \qquad (6.4)$$

Let us assume that the growth functions are linear functions of time:

$$\begin{aligned} g(a_l, \tau) &= A_l \tau + B_l \\ g(a_r, \tau) &= A_r \tau + B_r \end{aligned} \qquad (6.5)$$

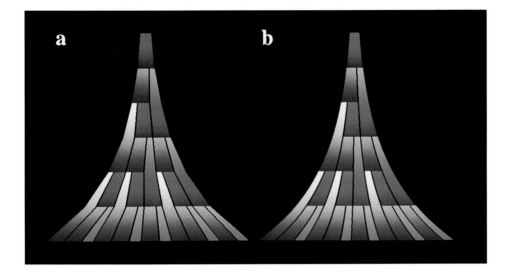

Figure 6.3: Diagrammatic representation of the development of *Anabaena catenula*, with (a) linear and (b) exponential growth of cells

By substituting equations (6.5) into (6.4), we obtain

$$
\begin{aligned}
2A_r + B_r &= (1A_l + B_l) + (0A_r + B_r) \quad \text{or} \quad 2A_r &= A_l + B_l \\
2A_l + B_l &= (0A_l + B_l) + (1A_r + B_r) \quad \text{or} \quad 2A_l &= A_r + B_r.
\end{aligned}
$$

The desired continuity of development is provided by all solutions of this system. They can be expressed in terms of coefficient c, which relates the growth rate of cells a_l to that of cells a_r:

$$
\begin{aligned}
A_l &= cA_r \\
B_l &= A_r(2 - c) \\
B_r &= A_r(2c - 1)
\end{aligned}
$$

Figure 6.3a illustrates the developmental process considered for $c = 1$. The diagram is obtained by plotting the cells in the filament as horizontal line segments on the screen. Colors indicate cell type and age. The observation time t ranges from 1 (at the top) to 7 (at the bottom), with increment $\Delta t = 0.009$.

The slopes of the side edges of the diagram represent growth rates of the entire structure. Notice that they remain constant in the periods between cell divisions, then change. This effect is disconcerting in animation, since the rate of organism growth suddenly increases with each cell division. In order to prevent this, it is necessary to extend requirements **R1** and **R2** to a higher order of continuity N. Specifically, equation (6.3) takes the form

$$
g^{(k)}(a, \beta) = \sum_{i=1}^{n} g^{(k)}(b_i, \alpha_i) \quad \text{for} \quad k = 0, 1, \ldots, N, \tag{6.6}
$$

where $g^{(k)}(a, \tau)$ is the kth derivative of the growth function $g(a, \tau)$ with respect to age τ.

Exponential growth

In the case of *Anabaena*, an attempt to achieve first-order continuity assuming linear growth functions yields an uninteresting solution, $g(a_l, \tau) = g(a_r, \tau) \equiv 0$. Thus, more complex growth functions must be considered, such as an exponential function that can be used to approximate the initial phase of sigmoidal growth. Assume that the growth function has the form

$$g(a_l, \tau) = g(a_r, \tau) = Ae^{B\tau}. \tag{6.7}$$

The objective is to find values of parameters A and B that satisfy equation (6.6) for $k = 0, 1$. By substituting equation (6.7) into (6.6), we obtain

$$AB^k e^{2B} = AB^k e^B + AB^k, \tag{6.8}$$

which implies that zero-order continuity yields continuity of infinite order in this case. Solution of equation (6.8) for any value of k yields

$$B = \ln \frac{1 + \sqrt{5}}{2} \approx 0.4812. \tag{6.9}$$

Parameter A is a scaling factor and can be chosen arbitrarily. The corresponding diagrammatic representation of development is shown in Figure 6.3b. The side edges of the diagram, representing the growth rates of the whole structure, are smooth exponential curves.

6.2.2   Development of branching structures

The notions of tDOL-system and growth function extend in a straightforward way to L-systems with brackets. For example, the following tDOL-system describes the recursive structure of the compound leaves analyzed in Section 5.3.

$$
\begin{aligned}
\omega \; : \quad & (a, 0) \\
p_1 \; : \quad & (a, 1) \quad \rightarrow \quad (s, 0)[(b, 0)][(b, 0)](a, 0) \\
p_2 \; : \quad & (b, \beta) \quad \rightarrow \quad (a, 0)
\end{aligned}
$$

According to production p_1, apex a produces an internode s, two lateral apices b and a younger apex a. Production p_2 transforms the lateral apices b into a after a delay β. The daughter branches recursively repeat the development of the mother branch.

Let us assume that the leaf development is first-order continuous, yielding the following equations for $k = 0, 1$:

$$
\begin{aligned}
g^{(k)}(a, 1) &= g^{(k)}(s, 0) + g^{(k)}(a, 0) & (6.10) \\
g^{(k)}(b, 0) &= 0 & (6.11) \\
g^{(k)}(b, \beta) &= g^{(k)}(a, 0) & (6.12)
\end{aligned}
$$

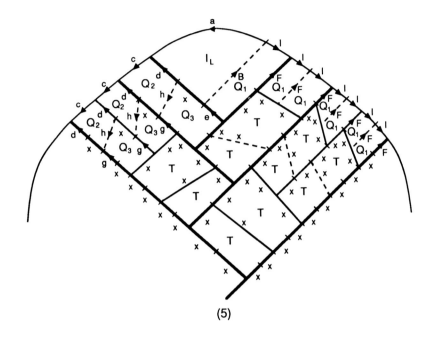

(5)

ω : $\vec{A}\overleftarrow{D}x\vec{b}$

l_1 : $\vec{a}$ $\rightarrow$ $\overleftarrow{A}\left[+\overleftarrow{b}\right]\vec{i}$ r_1 : $\vec{A}$ $\rightarrow$ $\overleftarrow{a}\left[-\overleftarrow{B}\right]\vec{I}$

l_2 : $\vec{b}$ $\rightarrow$ $\vec{e}\left[-\vec{B}\right]x\left[+\vec{h}\right]\vec{d}$ r_2 : $\vec{B}$ $\rightarrow$ $\vec{E}\left[+\vec{b}\right]x\left[-\vec{H}\right]\vec{D}$

l_3 : $\vec{d}$ $\rightarrow$ $\vec{f}$ r_3 : $\vec{D}$ $\rightarrow$ $\vec{F}$

l_4 : $\vec{f}$ $\rightarrow$ $\vec{g}\left[-\overleftarrow{h}\right]x\left[+\vec{h}\right]\vec{d}$ r_4 : $\vec{F}$ $\rightarrow$ $\vec{G}\left[+\vec{H}\right]x\left[-\vec{H}\right]\vec{D}$

l_5 : $\vec{h}$ $\rightarrow$ $x\left[-\vec{f}\right]x$ r_5 : $\vec{H}$ $\rightarrow$ $x\left[+\vec{F}\right]x$

l_6 : $\vec{i}$ $\rightarrow$ $\vec{c}$ r_6 : $\vec{I}$ $\rightarrow$ $\vec{C}$

l_7 : $\vec{c}$ $\rightarrow$ $\vec{i}\left[+\overleftarrow{f}\right]\vec{i}$ r_7 : $\vec{C}$ $\rightarrow$ $\vec{I}\left[-\overleftarrow{F}\right]\vec{I}$

l_8 : $\vec{e}$ $\rightarrow$ $x\left[+x\right]x$ r_8 : $\vec{E}$ $\rightarrow$ $x\left[-x\right]x$

l_9 : $\vec{g}$ $\rightarrow$ $x\left[-x\right]x\left[+x\right]x$ r_9 : $\vec{G}$ $\rightarrow$ $x\left[+x\right]x\left[-x\right]x$

Figure 7.10 (continued): Developmental sequence of *Microsorium*

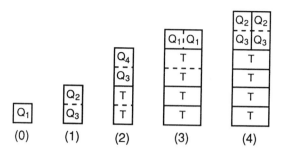

Figure 7.15: Developmental sequence of a *Dryopteris* segment

7.4   Dryopteris thelypteris

Gametophytes of other fern species follow a similar developmental pattern, with the apex producing segments alternately to the left and right. However, the cell division patterns within segments vary between species yielding different thalli shapes. For example, Figure 7.15 shows segment development in *Dryopteris thelypteris*. The corresponding cell division system is given below.

$$Q_1 \to \frac{Q_2}{Q_3} \qquad Q_2 \to \frac{Q_4}{Q_3} \qquad Q_3 \to \frac{T}{T} \qquad Q_4 \to Q_1 \mid Q_1$$

A developmental cycle of length 3, starting at cell Q_1, produces two new cells Q_1 separated by an anticlinal wall and a sequence of four terminal cells T separated by periclinal walls. A developmental sequence that combines the activity of the apex and the segments is shown in Figure 7.17. As in the case of the map L-system in Figure 7.12, only productions for the right side of the thallus are shown.

Map L-system   Apical cell division results from the application of productions r_1 and l_2 (creation of a right segment) or l_1 and r_2 (creation of a left segment). The subsequent cell divisions proceed in a symmetric way in right and left segments. The development of a right segment is described below.

The insertion of wall segment B creates the first cell Q_1 of segment $S_R^{(1)}$ (cf. Figure 7.8). Concurrently, wall D on the opposite side of the segment is transformed by production r_6 into FG (step 1). This transformation introduces a one-step delay before the application of production r_7 which, together with r_2, splits cell Q_1 into Q_2 and Q_3 by the first periclinal wall x (step 2). Meanwhile, production r_3 replaces wall C by wall J (step 1), after which r_4 replaces J by E (step 2). This introduces a two-step delay before cell Q_3 is subdivided periclinally into two cells T by production r_5 (step 3). In the same step, productions r_6 and r_9 subdivide cell Q_2 into cells Q_3 and Q_4, separated by periclinal wall H. Wall O from step 1 is transformed into wall R by productions

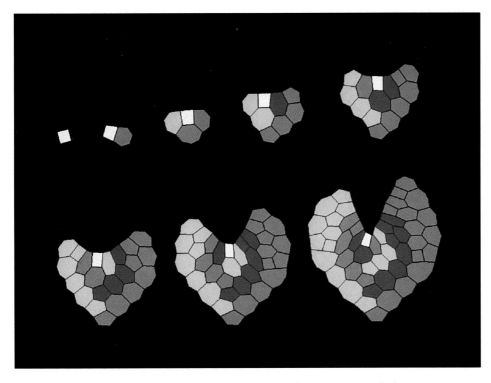

Figure 7.16: Simulated development of *Dryopteris thelypteris*

r_{15} (step 2) and r_{16} (step 3). Walls R and H are replaced by r_{17} and r_{14}, resulting in the first anticlinal division of cell Q_4 into two cells Q_1 by wall I (step 4). At the same time, productions r_5 and r_6 split cell Q_3 periclinally into two cells T. In the following derivation steps, each of the newly created cells Q_1 undergoes a sequence of changes similar to that described above. Production r_8 introduces a one-step delay before Q_1 is subdivided into Q_2 and Q_3 using r_9 and r_{10} (analogous to r_2 and r_7). Productions r_{11} and r_{12} play a role similar to r_5 and r_6, while production r_{13} introduces a delay. Walls labeled x do not undergo further changes and cells T do not subdivide. A simulated developmental sequence generated by the L-system in Figure 7.17 using the dynamic method to determine cell shape is given in Figure 7.16.

Microsorium vs. Dryopteris

A comparison of *Microsorium* and *Dryopteris* gametophytes (either real or modeled) indicates that different division patterns of segment cells have a large impact on the overall thalli shapes. In *Microsorium*, the number of *marginal cells*, situated at the apical front of a segment, is doubled every second derivation step. The segments are approximately triangular, with a wide apical front, which results in the circular thallus shape. The apical front of *Dryopteris* segments is comparatively less developed. The number of marginal cells is doubled only every third step, and the segments grow faster in length. The resulting thallus shape is concave.

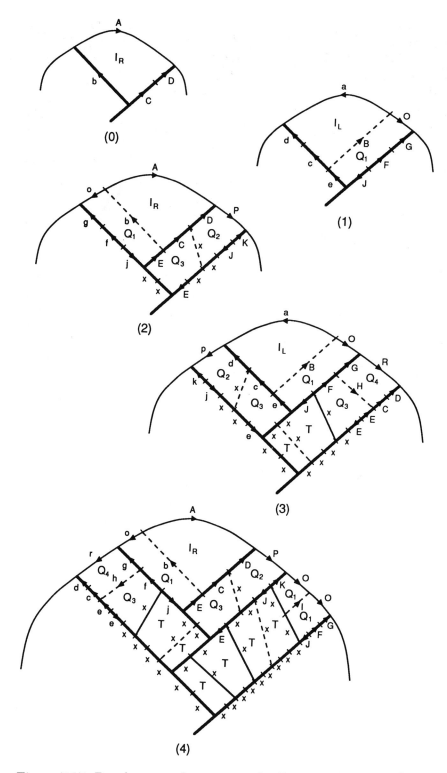

Figure 7.17: Developmental sequence of a *Dryopteris* gametophyte

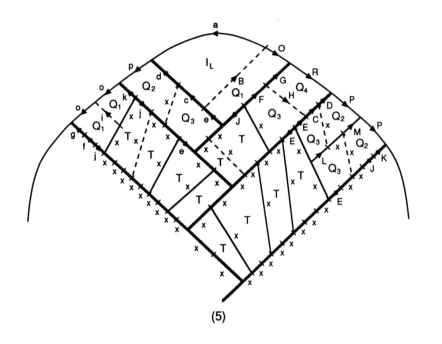

(5)

$$\omega : \quad \overrightarrow{A}\overleftarrow{D}\overleftarrow{C}\overrightarrow{b}$$

$$r_1 : \quad \overrightarrow{A} \quad \rightarrow \quad \overleftarrow{a}\left[-\overleftarrow{B}\right]\overrightarrow{O}$$

$$r_2 : \quad \overrightarrow{B} \quad \rightarrow \quad \overrightarrow{E}\left[+\overleftarrow{b}\right]\overleftarrow{C}\left[-x\right]\overrightarrow{D}$$

$$r_3 : \quad \overrightarrow{C} \quad \rightarrow \quad \overleftarrow{J}$$

$$r_4 : \quad \overrightarrow{J} \quad \rightarrow \quad \overrightarrow{E}$$

$$r_5 : \quad \overrightarrow{E} \quad \rightarrow \quad x\left[-x\right]x$$

$$r_6 : \quad \overrightarrow{D} \quad \rightarrow \quad \overrightarrow{F}\left[-\overrightarrow{H}\right]\overrightarrow{G}$$

$$r_7 : \quad \overrightarrow{F} \quad \rightarrow \quad x\left[+x\right]x\left[-x\right]\overleftarrow{J}$$

$$r_8 : \quad \overrightarrow{G} \quad \rightarrow \quad \overrightarrow{K}$$

$$r_9 : \quad \overrightarrow{K} \quad \rightarrow \quad \overrightarrow{E}\left[+\overleftarrow{H}\right]\overleftarrow{C}\left[-x\right]\overrightarrow{D}$$

$$r_{10} : \quad \overrightarrow{I} \quad \rightarrow \quad \overrightarrow{L}\left[+x\right]x\left[-x\right]\overrightarrow{M}$$

$$r_{11} : \quad \overrightarrow{L} \quad \rightarrow \quad x\left[+x\right]x\left[-x\right]x$$

$$r_{12} : \quad \overrightarrow{M} \quad \rightarrow \quad \overrightarrow{L}\left[+\overleftarrow{H}\right]x\left[-\overrightarrow{H}\right]\overrightarrow{N}$$

$$r_{13} : \quad \overrightarrow{N} \quad \rightarrow \quad \overrightarrow{I}$$

$$r_{14} : \quad \overrightarrow{H} \quad \rightarrow \quad x\left[+\overrightarrow{I}\right]x$$

$$r_{15} : \quad \overrightarrow{O} \quad \rightarrow \quad \overrightarrow{P}$$

$$r_{16} : \quad \overrightarrow{P} \quad \rightarrow \quad \overrightarrow{Q}$$

$$r_{17} : \quad \overrightarrow{Q} \quad \rightarrow \quad \overrightarrow{O}\left[-\overleftarrow{I}\right]\overrightarrow{O}$$

Figure 7.17 (continued): Developmental sequence of *Dryopteris*

In order to quantify the relationship between segment cell division pattern and thallus shape, de Boer [22] proposed an empirical measure called *periclinal ratio*. It is based on the following considerations:

- anticlinal growth takes place mainly along the margin and is exponential since all marginal cells divide; and

- periclinal growth of a segment is linear, as cells displaced away from the margin eventually stop dividing.

Since the division pattern is recursive, the average ratio of the numbers of marginal cells in neighboring segments converges to a constant A. Similarly, the difference between the numbers of cells along the periclinal boundary of two neighboring segments converges to a constant P. By computer simulation [22], it was found that if the periclinal ratio P/A is smaller than 1.25, the apical front is convex as in the case of *Microsorium linguaeforme*. In contrast, if P/A is larger than 2.0, the simulated structure develops a concave apical front, corresponding to a heart-shaped thallus as in *Dryopteris thelypteris*. These results relate local growth to global shape. Periclinal ratios from 1.25 to 2.0 were not studied in detail.

7.5 Modeling spherical cell layers

Animal embryos

Although the scope of this book is limited to plants, it is interesting to note that the formalism of L-systems can be applied also to simulate some developmental processes in animals. For example, during the *cleavage* stage of development, an animal embryo consists of a single layer of cells that covers the surface of a spherical cavity. This structure is known as the *blastula* [6]. The cells divide synchronously in a regular pattern up to and including the 64-cell stage (6th cleavage). This development can be captured using an mBPMOL-system operating on the surface of a sphere rather than on a plane. To this end, cell walls are represented as great circle arcs connecting vertices that are constrained to the sphere surface.

The extension of the dynamic interpretation method from the plane to the surface of a sphere requires few changes. Osmotic pressure and wall tension are calculated as before. Since the resulting force may displace a vertex away from the surface of the sphere, the actual vertex position is found by projecting the displaced point back to the sphere. During the cleavage stage, cells of embryos do not expand, thus the overall size of the sphere is constant.

Patella vulgata

For example, deBoer [22] proposed the map L-system in Figure 7.18 to model the development of a snail embryo, *Patella vulgata*, according to data presented by Biggelaar [150]. The starting map and developmental sequence are shown in Figure 7.18, while Figure 7.19 presents an alternative rendering. Each cell is approximated by a sphere centered at the point obtained by raising the center of gravity of the cell

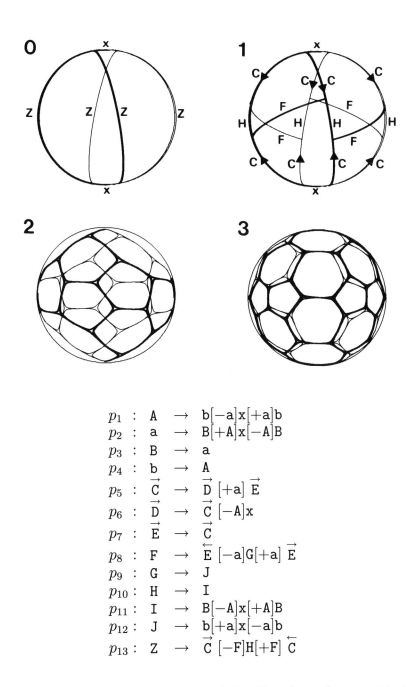

$$
\begin{aligned}
p_1 &: \quad \text{A} \quad \rightarrow \quad \text{b[}-\text{a]x[}+\text{a]b} \\
p_2 &: \quad \text{a} \quad \rightarrow \quad \text{B[}+\text{A]x[}-\text{A]B} \\
p_3 &: \quad \text{B} \quad \rightarrow \quad \text{a} \\
p_4 &: \quad \text{b} \quad \rightarrow \quad \text{A} \\
p_5 &: \quad \overrightarrow{\text{C}} \quad \rightarrow \quad \overrightarrow{\text{D}} \text{ [}+\text{a] } \overrightarrow{\text{E}} \\
p_6 &: \quad \overrightarrow{\text{D}} \quad \rightarrow \quad \overrightarrow{\text{C}} \text{ [}-\text{A]x} \\
p_7 &: \quad \overrightarrow{\text{E}} \quad \rightarrow \quad \overrightarrow{\text{C}} \\
p_8 &: \quad \text{F} \quad \rightarrow \quad \overleftarrow{\text{E}} \text{ [}-\text{a]G[}+\text{a] } \overrightarrow{\text{E}} \\
p_9 &: \quad \text{G} \quad \rightarrow \quad \text{J} \\
p_{10} &: \quad \text{H} \quad \rightarrow \quad \text{I} \\
p_{11} &: \quad \text{I} \quad \rightarrow \quad \text{B[}-\text{A]x[}+\text{A]B} \\
p_{12} &: \quad \text{J} \quad \rightarrow \quad \text{b[}+\text{a]x[}-\text{a]b} \\
p_{13} &: \quad \text{Z} \quad \rightarrow \quad \overrightarrow{\text{C}} \text{ [}-\text{F]H[}+\text{F] } \overleftarrow{\text{C}}
\end{aligned}
$$

Figure 7.18: Developmental sequence of *Patella vulgata* (equatorial view)

vertices to the surface of the underlying spherical cavity. The radius is the maximum distance from the center of gravity to the cell vertices. A comparison of the *Patella* model at the 16-cell stage (bottom left of Figure 7.19) with an electron microscope image (Figure 7.20) shows good correspondence between the model and reality.

7.6 Modeling 3D cellular structures

Cellworks

The previous sections presented a method for modeling cellular layers extending in a plane or on the surface of a sphere. However, real cellular structures are three-dimensional objects. In order to capture the three-dimensional aspect of cellular layers and model more complex structures, Lindenmayer [85] proposed an extension of map L-systems called *cellwork* L-systems. The notion of a cellwork is characterized as follows.

- A cellwork is a finite set of *cells*. Each cell is surrounded by one or more *walls* (faces).

- Each wall is surrounded by a boundary consisting of a finite, circular sequence of *edges* that meet at *vertices*.

- Walls cannot intersect without forming an edge, although there can be walls without edges (in the case of cells shaped as spheres or tori).

- Every wall is part of the boundary of a cell, and the set of walls is connected.

- Each edge has one or two vertices associated with it. The edges cannot cross without forming a vertex, and there are no vertices without an associated edge.

- Every edge is a part of the boundary of a wall, and the set of edges is connected.

Note that the terms *cell* and *wall* have different meanings in the context of cellworks than in the context of maps.

mBPCOL-system

The development of three-dimensional structures is captured using an extension of mBPMOL-systems called *marker Binary Propagating Cellwork OL-systems* [42]. An mBPCOL-system $\mathcal{G}$ is defined by a finite alphabet of *edge labels* Σ, a finite alphabet of *wall labels* Γ, a *starting cellwork* ω, and a finite set of *edge productions* P. The initial cellwork ω is specified by a list of walls and their bounding edges. As in the case of mBPMOL-systems, edges may be directed or neutral. Each production is of the form $A : \beta \to \alpha$, where the edge $A \in \Sigma$ is the *predecessor*, the string $\beta \in \{\Gamma^+, *\}$ is a list of *applicable walls* (* denotes all walls), and the string α is the *successor*, which is composed of edge labels from Σ,

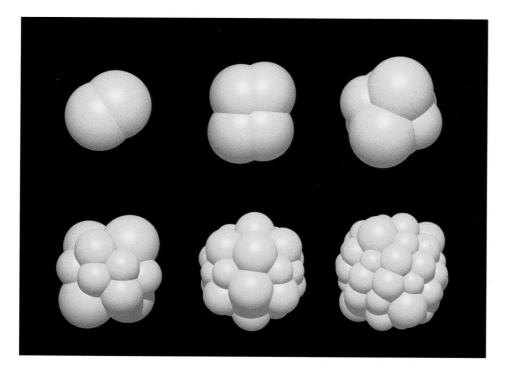

Figure 7.19: Simulated development of *Patella vulgata*

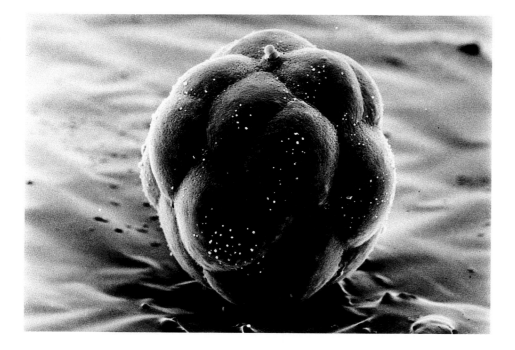

Figure 7.20: An electron microscope image of *Patella vulgata*

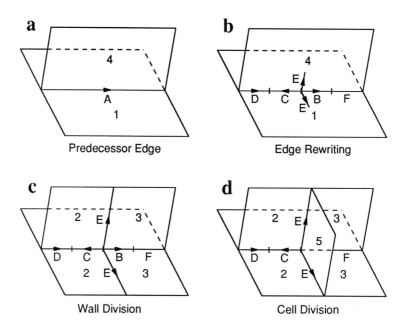

Figure 7.21: The phases of a derivation step

wall labels from Γ and the symbols [and]. The symbols outside square brackets describe the subdivision pattern for the predecessor. Pairs of matching brackets [and] delimit *markers* that specify possible attachment sites for new edges and walls. As in the two-dimensional case, arrows indicate the relative directions of successor edges and markers with respect to the predecessor edge. The list β contains all walls into which a marker should be inserted. In addition to the labels for edges and markers, a production successor specifies the labels of walls that may be created as a result of a derivation step.

Production syntax

The syntax of a production is best explained using an example. The production

$$\vec{A} : 14 \rightarrow \vec{D}\vec{C}_2[\overleftarrow{E}_5]\vec{B}_3\vec{F}$$

applies to an edge A if it belongs to one or more walls labeled 1 or 4 (Figure 7.21a). The predecessor edge is subdivided into four edges D, C, B and F. During a derivation step, marker E is introduced into all walls of type 1 or 4 that share edge A (Figure 7.21b), and can be connected with a matching marker inserted into the same wall by another production. As a result, the wall will split into two. The daughter wall having C as one of its edges will be labeled 2, and the wall having B as an edge will be labeled 3 (Figure 7.21c). Markers E can be connected only if both productions assign labels to the daughter walls in a consistent way. Otherwise, the markers are considered nonmatching and are discarded. If several walls bounding a cell split in such a way that the sequence of new edges forms a closed contour, a

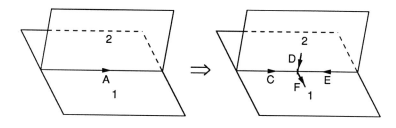

Figure 7.22: Example of consistent edge productions

new wall bound by these edges may be created. In order for this to occur, all markers involved must specify the same label for the new wall, 5 in this example (Figure 7.21d).

The limitation of the scope of a production to specific walls may create a consistency problem while rewriting edges. For instance, assume that walls 1 and 2 share edge A and the following productions are in P:

$$p_1 : \quad \overrightarrow{A} : 1 \quad \rightarrow \quad \overrightarrow{C}\overleftarrow{E}$$
$$p_2 : \quad \overrightarrow{A} : 2 \quad \rightarrow \quad \overrightarrow{A}\overrightarrow{B}$$

Productions p_1 and p_2 are inconsistent since they specify two different partitions of the same edge. It is assumed that the mBPCOL-systems under consideration are free of such inconsistencies. This does not preclude the possibility of applying several productions simultaneously to the same edge. For example, a production pair,

$$p_1 : \quad \overrightarrow{A} : 1 \quad \rightarrow \quad \overrightarrow{C}_2[\overrightarrow{F}_3]\overleftarrow{E}_4$$
$$p_2 : \quad \overrightarrow{A} : 2 \quad \rightarrow \quad \overrightarrow{C}_5[\overleftarrow{D}_6]\overleftarrow{E}_7,$$

consistently divides edge A into segments C and E, although the markers inserted into walls 1 and 2 are different (Figure 7.22).

According to the above discussion, a *derivation step* in an mBPCOL- system consists of three phases.

Derivation

- Each edge in the cellwork is replaced by successor edges and markers using one or more productions in P.

- Each wall is scanned for matching markers. If a match inducing a consistent labeling of daughter cells is found, the wall is subdivided. The selection of matching markers is done by the system. Unused markers are discarded.

- Each cell is scanned for a circular sequence of new division edges. If a cycle assigning the same label to the division wall is found, it is used to bound the wall that will divide the cell into two daughter cells. If different possibilities exist, the edges are selected by the system.

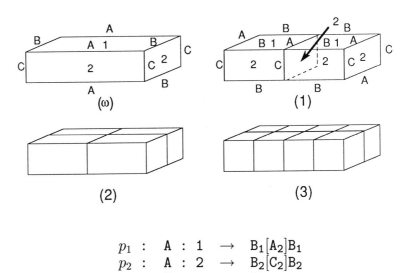

$$
\begin{array}{llll}
p_1 : & A : 1 & \rightarrow & B_1[A_2]B_1 \\
p_2 : & A : 2 & \rightarrow & B_2[C_2]B_2 \\
p_3 : & B : * & \rightarrow & A
\end{array}
$$

Figure 7.23: Example of a cellwork L-system

A wall may be subdivided more than once as long as new division edges do not intersect and a consistent labeling of daughter walls is possible. In contrast, a cell may be divided only once in any derivation step.

Example

For example, Figure 7.23 presents a three-dimensional extension of the map L-system from Figure 7.2. In the first derivation step, production p_1 divides walls labeled 1, and production p_2 divides walls labeled 2. The inserted edges form a cycle that divides the cell with a new wall labeled 2. In the subsequent steps this process is repeated, generating a pattern of alternating division walls. Production p_3 introduces the necessary delay.

Dynamic interpretation

The dynamic method for interpreting map L-systems is extended to cellwork L-systems using the following assumptions:

- the structure is represented as a three-dimensional network of masses corresponding to cell vertices, connected by springs which correspond to cell edges,

- the springs are always straight and obey Hooke's law,

- for the purpose of force calculations, walls can be approximated by flat polygons,

- the cells exert pressure on their bounding walls; the pressure on a wall is directly proportional to the wall area and inversely proportional to the cell volume,

- the pressure on a wall is divided evenly between the wall vertices,

- the motion of masses is damped, and

- other forces are not considered.

The total force $\vec{F}_T$ acting on a vertex X is given by the formula

$$\vec{F}_T = \sum_{e \in E} \vec{F}_e + \sum_{w \in W} \vec{F}_w + \vec{F}_d,$$

where $\vec{F}_e$ are forces contributed by the set of edges E incident to X, $\vec{F}_w$ are forces contributed by the set of walls W incident to X, and $\vec{F}_d = -b\vec{v}$ is a damping force. The forces $\vec{F}_e$ act along the cell edges and represent wall *tension*. The forces $\vec{F}_w$ are due to the *pressure* exerted by the cells on their bounding walls. The total force of pressure $\vec{P}$ exerted by a cell on a wall w has direction normal to w and is equal to $p \cdot A$, where p is the internal cell pressure and A is the wall area. Calculation of the polygon area proceeds as in the two-dimensional case. The pressure p is assumed to be inversely proportional to the cell volume, $p \sim V^{-1}$, which corresponds to the equation describing osmotic pressure (Section 7.2). The volume V of a cell is calculated by tesselating the cell into tetrahedra. The resulting differential equations are formed and solved as in the two-dimensional case.

A division pattern that frequently occurs in epidermal cell structures *Epidermal cells* is described by the L-system in Figure 7.24, based on a cyclic cellwork L-system (a slightly different formalism) proposed by Lindenmayer [85]. Productions p_1, p_2, p_6 and p_7 are responsible for cell divisions, while the remaining productions introduce delays such that the division pattern is staggered.

On the surface, the cellular structures analyzed in this chapter may appear quite unrelated to the models discussed previously. However, a closer inspection reveals many analogies. For example, consecutively created segments of a fern gametophyte exhibit a phase effect corresponding to that observed in inflorescences. Furthermore, parts of an older gametophyte situated near the apex have the same topology as the entire thallus at an earlier developmental stage, which associates the recursive structures generated by map or cellwork L-systems with self-similar patterns created using string L-systems. As observed by Oppenheimer [105], self-similarity appears to be one of the general principles organizing the world of botany. The next chapter discusses this topic in more detail.

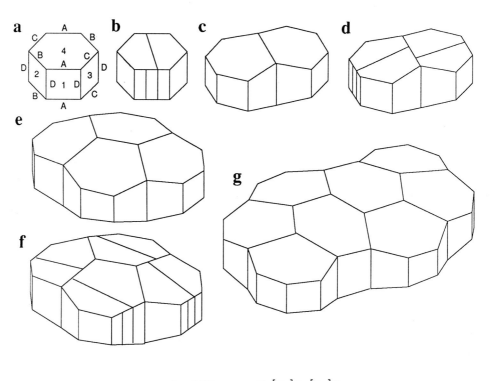

$$
\begin{array}{llll}
p_1: & \text{A}:123 & \rightarrow & \text{C}_3[\text{E}_1]\text{B}_2[\text{D}_1]\text{C}_3 \\
p_2: & \text{A}:4 & \rightarrow & \text{CB}_4[\text{F}_1]\text{C}_4 \\
p_3: & \text{B}:* & \rightarrow & \text{A} \\
p_4: & \text{C}:* & \rightarrow & \text{B} \\
p_5: & \text{E}:* & \rightarrow & \text{D} \\
p_6: & \text{F}:123 & \rightarrow & \text{HGH} \\
p_7: & \text{F}:4 & \rightarrow & \text{H}_4[\text{F}_1]\text{G}_4[\text{F}_1]\text{H}_4 \\
p_8: & \text{G}:* & \rightarrow & \text{F} \\
p_9: & \text{H}:* & \rightarrow & \text{G}
\end{array}
$$

Figure 7.24: Developmental sequence of epidermal cells: (a) The starting cellwork; (b), (d) and (f) cellworks immediately after cell divisions; (c), (e) and (g) the corresponding cellworks at equilibrium

Chapter 8

Fractal properties of plants

What is a fractal? In his 1982 book, Mandelbrot defines it as a set with *Fractals vs.*
Hausdorff-Besicovitch dimension D_H strictly exceeding the topological *finite curves*
dimension D_T [95, page 15]. In this sense, none of the figures presented
in this book are fractals, since they all consist of a finite number of
primitives (lines or polygons), and $D_H = D_T$. However, the situation
changes dramatically if the term "fractal" is used in a broader sense [95,
page 39]:

> Strictly speaking, the triangle, the Star of David, and the
> finite Koch teragons are of dimension 1. However, both
> intuitively and from the pragmatic point of view of the sim-
> plicity and naturalness of the corrective terms required, it is
> reasonable to consider an advanced Koch teragon as being
> closer to a curve of dimension log 4/log 3 than to a curve
> of dimension 1.

Thus, a finite curve can be considered an approximate rendering
of an infinite fractal as long as the interesting properties of both are
closely related. In the case of plant models, this distinctive feature is
self-similarity.

The use of approximate figures to illustrate abstract concepts has a *Fractals vs.*
long tradition in geometry. After all, even the primitives of Euclidean *plants*
geometry — a point and a line — cannot be *drawn* exactly. An interesting question, however, concerns the relationship between fractals
and real biological structures. The latter consist of a finite number of
cells, thus are not fractals in the strict sense of the word. To consider
real plants as approximations of "perfect" fractal structures would be
acceptable only if we assumed Plato's view of the supremacy of ideas
over their mundane realization. A viable approach is the opposite one,
to consider fractals as abstract descriptions of the real structures. At
first sight, this concept may seem strange. What can be gained by *Complexity of*
reducing an irregular contour of a compound leaf to an even more ir- *fractals*
regular fractal? Would it not be simpler to characterize the leaf using

a smooth curve? The key to the answer lies in the meaning of the term
"simple." A smooth curve may seem intuitively simpler than a fractal,
but as a matter of fact, the reverse is often true [95, page 41]. According to Kolmogorov [80], the complexity of an object can be measured
by the length of the shortest algorithm that generates it. In this sense,
many fractals are particularly simple objects.

Previous
viewpoints

The above discussion of the relationship between fractals and plants
did not emerge in a vacuum. Mandelbrot [95] gives examples of the recursive branching structures of trees and flowers, analyzes their
Hausdorff-Besicovitch dimension and writes inconclusively "trees may
be called fractals in part." Smith [136] recognizes similarities between
algorithms yielding Koch curves and branching plant-like structures,
but does not qualify plant models as fractals. These structures are produced in a finite number of steps and consist of a finite number of line
segments, while the "notion of fractal is defined only in the limit." Oppenheimer [105] uses the term "fractal" more freely, exchanging it with
self-similarity, and comments: "The geometric notion of self-similarity
became a paradigm for structure in the natural world. Nowhere is this
principle more evident than in the world of botany." The approach presented in this chapter, which considers fractals as simplified abstract
representations of real plant structures, seems to reconcile these previous opinions.

Fractals in
botany

But why are we concerned with this problem at all? Does the notion of fractals provide any real assistance in the analysis and modeling
of real botanical structures? On the conceptual level, the distinctive
feature of the fractal approach to plant analysis is the emphasis on
self-similarity. It offers a key to the understanding of complex-looking,
compound structures, and suggests the recursive developmental mechanisms through which these structures could have been created. The
reference to similarities in living structures plays a role analogous to the
reference to symmetry in physics, where a strong link between conser-
vation laws and the invariance under various symmetry operations can
be observed. Weyl [159, page 145] advocates the search for symmetry
as a cognitive tool:

> Whenever you have to deal with a structure-endowed entity
> Σ, try to determine its group of automorphisms, the group
> of those element-wise transformations which leave all struc-
> tural relations undisturbed. You can expect to gain a deep
> insight into the constitution of Σ in this way.

The relationship between symmetry and self-similarity is discussed
in Section 8.1. Technically, the recognition of self-similar features of
plant structures makes it possible to render them using algorithms developed for fractals as discussed in Section 8.2.

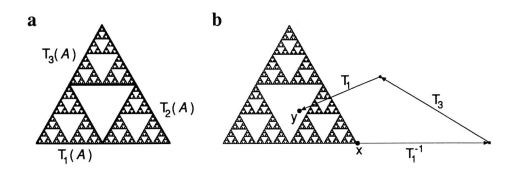

Figure 8.1: The Sierpiński gasket is closed with respect to transformations T_1, T_2 and T_3 (a), but it is not closed with respect to the set including the inverse transformations (b).

8.1   Symmetry and self-similarity

The notion of symmetry is generally defined as the invariance of a configuration of elements under a group of automorphic transformations. Commonly considered transformations are congruences, which can be obtained by composing rotations, reflections and translations. Could we extend this list of transformations to similarities, and consider self-similarity as a special case of symmetry involving scaling operations?

On the surface, this seems possible. For example, Weyl [159, page 68] suggests: "In dealing with potentially infinite patterns like band ornaments or with infinite groups, the operation under which a pattern is invariant is not of necessity a congruence but could be a similarity." The spiral shapes of the shells *Turritella duplicata* and *Nautilus* are given as examples. However, all similarities involved have the same fixed point. The situation changes dramatically when similarities with different fixed points are considered. For example, the Sierpiński gasket is mapped onto itself by a set of three contractions T_1, T_2 and T_3 (Figure 8.1a). Each contraction takes the entire figure into one of its three main components. Thus, if A is an arbitrary point of the gasket, and $T = T_{i_1} T_{i_2} \ldots T_{i_n}$ is an arbitrary composition of transformations T_1, T_2 and T_3, the image $T(A)$ will belong to the set A. On the other hand, if the inverses of transformations T_1, T_2 and T_3 can also be included in the composition, one obtains points that do not belong to the set A nor its infinite extension (Figure 8.1b). This indicates that the set of transformations that maps A into itself forms a semigroup generated by T_1, T_2 and T_3, but does not form a group. Thus, self-similarity is a weaker property than symmetry, yet it still provides a valuable insight into the relationships between the elements of a structure.

Figure 8.2: The fern leaf from Barnsley's model [7]

8.2   Plant models and iterated function systems

Barnsley [7, pages 101–104] presents a model of a fern leaf (Figure 8.2), generated using an *iterated function system*, or IFS. This raises a question regarding the relationship between developmental plant models expressed using L-systems and plant-like structures captured by IFSes. This section briefly describes IFSes and introduces a method for constructing those which approximate structures generated by a certain type of parametric L-system. The restrictions of this method are analyzed, shedding light on the role of IFSes in the modeling of biological structures.

IFS definition   By definition [74], a planar iterated function system is a finite set of contractive affine mappings $\mathcal{T} = \{T_1, T_2, \ldots, T_n\}$ which map the plane $\mathcal{R} \times \mathcal{R}$ into itself. The *set defined by $\mathcal{T}$* is the smallest nonempty set $\mathcal{A}$, closed in the topological sense, such that the image y of any point $x \in \mathcal{A}$ under any of the mappings $T_i \in \mathcal{T}$ also belongs to $\mathcal{A}$. It can be shown that such a set always exists and is unique [74] (see also [118] for an elementary presentation of the proof). Thus, starting from an arbitrary point $x \in \mathcal{A}$, one can approximate $\mathcal{A}$ as a set of images of x under compositions of the transformations from $\mathcal{T}$. On

a   **b**   **c**

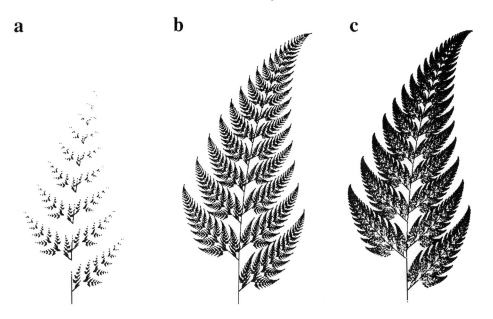

Figure 8.3: A comparison of three attracting methods for the rendering of a set defined by an IFS: (a) deterministic method using a balanced tree of depth $n = 9$ with the total number of points $N_1 = 349,525$, (b) deterministic method using a non-balanced tree with $N_2 = 198,541$ points, (c) stochastic method with $N_3 = N_2$ points

the other hand, if the starting point x does not belong to $\mathcal{A}$, the consecutive images of x gradually approach $\mathcal{A}$, since all mappings T_i are contractions. For this reason, the set $\mathcal{A}$ is called the *attractor* of the IFS $\mathcal{T}$. The methods for rendering it are based on finding the images $T_{i_k}(T_{i_{k-1}}(\ldots(T_{i_1}(x))\ldots)) = xT_{i_1}\ldots T_{i_{k-1}}T_{i_k}$, and are termed *attracting methods*. According to the *deterministic approach* [123], a tree of transformations is constructed, with each node representing a point in $\mathcal{A}$. Various strategies, such as breadth-first or depth-first, can be devised to traverse this tree and produce different intermediate results [60]. If the transformations in $\mathcal{T}$ do not have the same scaling factors (Lipschitz constants), the use of a balanced tree yields a non-uniform distribution of points in $\mathcal{A}$. This effect can be eliminated by constructing a non-balanced tree, using a proper criterion for stopping the extension of a branch [60]. An alternative approach for approximating the set $\mathcal{A}$ is termed the *chaos game* [7] (see also [107, Chapter 5]). In this case, only one sequence of transformations is constructed, corresponding to a single path in the potentially infinite tree of transformations. The transformation applied in each derivation step is selected at random. In order to achieve a uniform distribution of points in the attractor, the probability of choosing transformation $T_i \in \mathcal{T}$ is set according to its Lipschitz constant. Figure 8.3 illustrates the difference between the stochastic and deterministic methods of rendering the attractor. The

Rendering methods

underlying IFS consists of four transformations, given below using homogeneous coordinates [40]:

$$T_1 = \begin{bmatrix} 0.00 & 0.00 & 0.00 \\ 0.00 & 0.16 & 0.00 \\ 0.00 & 0.00 & 1.00 \end{bmatrix}$$

$$T_2 = \begin{bmatrix} 0.20 & 0.23 & 0.00 \\ -0.26 & 0.22 & 0.00 \\ 0.00 & 1.60 & 1.00 \end{bmatrix}$$

$$T_3 = \begin{bmatrix} -0.15 & 0.26 & 0.00 \\ 0.28 & 0.24 & 0.00 \\ 0.00 & 0.44 & 1.00 \end{bmatrix}$$

$$T_4 = \begin{bmatrix} 0.85 & -0.04 & 0.00 \\ 0.04 & 0.85 & 0.00 \\ 0.00 & 1.60 & 1.00 \end{bmatrix}$$

Other methods for the rendering of the set $\mathcal{A}$, defined by an interated function system $\mathcal{T}$, include the *repelling* or *escape-time* method and the *distance* method [60, 118]. Both methods assign values to points outside of $\mathcal{A}$. The first method determines how fast a point is repelled from $\mathcal{A}$ to infinity by the set of inverse transformations T_i^{-1}, where $T_i \in \mathcal{T}$. An example of the application of this method, with escape time values represented as a height field, is shown in Figure 8.4. The second method computes the Euclidean distance of a point from the attractor $\mathcal{A}$.

IFS construction

The problem of constructing an IFS that will approximate a branching structure modeled using an L-system can now be considered. This discussion focuses specifically on structures that develop in a biologically justifiable way, by subapical branching (Section 3.2). The compound leaf shown in Figure 5.11a on page 129 will be used as a working example. In this case, the apical delay D is equal to zero, and the L-system can be represented in the simplified form:

$$\begin{aligned} \omega &: A \\ p_1 &: A & : & * & \to & F(1)[+A][-A]F(1)A \\ p_2 &: F(a) & : & * & \to & F(a * R) \end{aligned} \qquad (8.1)$$

Figure 8.4: Fern dune $\longrightarrow$

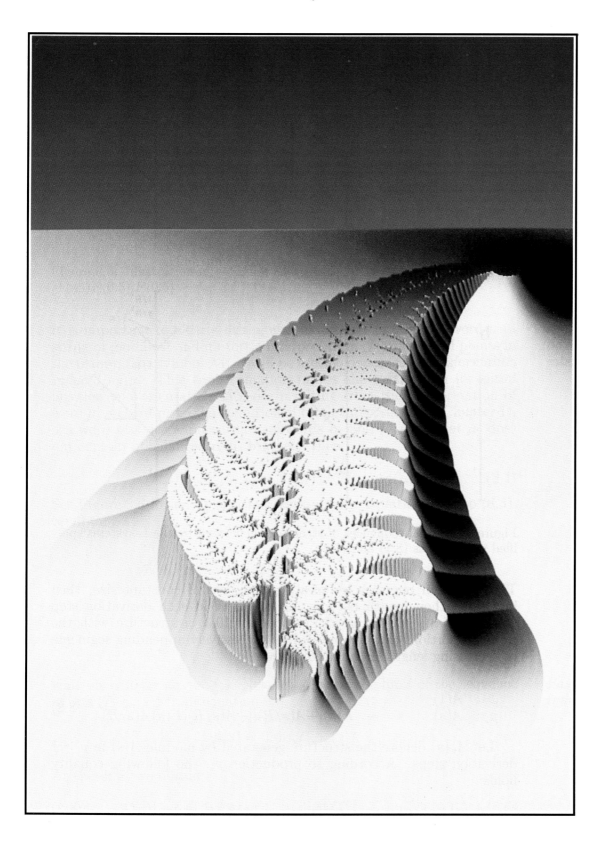

a   **b**

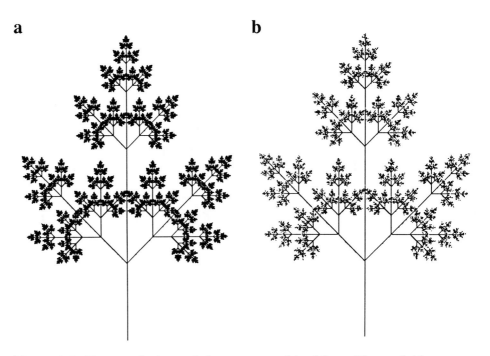

Figure 8.8: Two renderings of the compound leaf from Figure 5.11a, generated using iterated function systems

This last value is calculated as the limit of the geometric series with the first term equal to $2s$ and the ratio equal to $1/R$. Thus, the compound leaf of Figure 5.11a is defined by an IFS consisting of transformation

$$
Q = \begin{bmatrix} 0 & 0 & 0 \\ 0 & 1 - 1/R & 0 \\ 0 & 0 & 1 \end{bmatrix}
$$

and transformations T_1, T_2 and T_3 specified as in the case of the controlled IFS.

Rendering examples   Two fractal-based renderings of the set $\mathcal{A}(s)$ are shown in Figure 8.8. Figure 8.8a was obtained using the controlled IFS and a deterministic algorithm to traverse the tree of admissible transformations. Figure 8.8b was obtained using the "ordinary" IFS and the random selection of transformations. Figure 8.9 shows another fractal-based rendering of the same structure. The spheres have radii equal to the distance from the sphere center to the leaf, within a specified ϵ.

Figure 8.9: Carrot leaf $\longrightarrow$

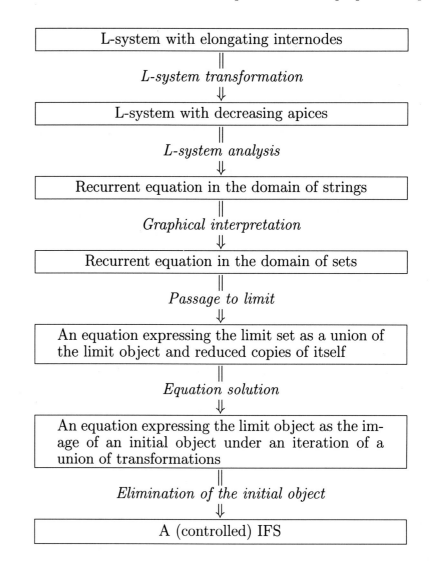

Figure 8.10: Steps in the construction of an IFS given an L-system capturing a developmental model

It is instructive to retrace the logical construction that started with an L-system, and ended with an iterated function system which can generate fractal approximations of the same object (Figure 8.10). An analysis of the operations performed in the subsequent steps of this construction reveals its limitations, and clarifies the relationship between strictly self-similar structures and real plants. The critical step is the transformation of the L-system with elongating internodes to the L-system with decreasing apices. It can be performed as indicated in the example if the plant maintains constant branching angles as well as fixed proportions between the mother and daughter segments, independent of branch order. This, in turn, can be achieved if all segments in the modeled plant elongate exponentially over time. These are strong assumptions, and may be satisfied to different degrees in real plants. Strict self-similarity is an abstraction that captures the essential properties of many plant structures and represents a useful point of reference when describing them in detail.

Conclusions

Epilogue

This quiet place, reminiscient of Claude Monet's 1899 painting *Water-lilies pool — Harmony in green*, does not really exist. The scene was modeled using L-systems that captured the development of trees and water plants, and illuminated by simulated sunlight. It is difficult not to appreciate how far the theory of L-systems and the entire field of computer graphics have developed since their beginnings in the 1960's, making such images possible. Yet the results contained in this book are not conclusive and constitute only an introduction to the research on plant modeling for biological and graphics purposes. The algorithmic beauty of plants is open to further exploration.

← Figure E.1: Water-lilies

Appendix A

Software environment for plant modeling

This book is illustrated with images of plants which exist only as mathematical models visualized by means of computer graphics. The software environment used to construct and experiment with these models includes dozens of programs and hundreds of data files. This creates the nontrivial problem of organizing all components for easy definition, saving, retrieval and modification of the models. In order to solve it, the idea of simulation was extended beyond the level of individual plants to an entire laboratory in botany [98]. Thus, a user can create and conduct experiments in a *virtual laboratory* by applying intuitive concepts and techniques from the "real" world. As an operating system defines the way a user perceives a computing environment, the virtual laboratory determines a user's perception of the environment in which simulated experiments take place. In the future, a virtual laboratory may complement, extend, or even replace books as a means for gathering and presenting scientific information. Because of this potential, the laboratory in which the research reported in this book was produced is described here in more detail.

A.1   A virtual laboratory in botany

A virtual laboratory, like its "real" counterpart, is a playground for experimentation. It comes with a set of *objects* pertinent to its scientific domain (in this case, plant models), *tools* which operate on these objects, a *reference book* and a *notebook*. Once the concepts and tools are understood, the user can expand the laboratory by adding new objects, creating new experiments, and recording descriptions in the notebook. An experienced user can expand the laboratory further by creating and installing new tools.

User's perspective

Laboratory =
microworld +
hypertext

Technically, a virtual laboratory is a *microworld* which can be explored under the guidance of a *hypertext* system. The term "microworld" denotes an interactive environment for creating and conducting simulated experiments. The guidance could be provided in the form of a traditional book, but an electronic document is more suitable for integration with a microworld. In a sense, both components of the virtual laboratory are described by Nelson in *Dream Machines* [104]. The pioneering role of this book in introducing the concept of hypertext is known, but under the heading *The Mind's Eye* the notion of a microworld is also anticipated:

> Suppose that you have a computer.
> What sorts of things would you do with it?
> Things that are imaginative
>   and don't require too much else.
> I am hinting at something.
> You could have it make pictures and show you stuff
> and change what it shows depending on what you do.

Requirements

A virtual laboratory can be divided into two components: the application programs, data files and textual descriptions that comprise the experiments; and the system support that provides the framework on which these domain-dependent experiments are built. The following list specifies the features of this framework.

- **Consistent organization of the lab.** In the lab environment, experiments are run by applying tools (programs) to objects (data files). An object consists of files that are grouped together so that they can be retrieved easily. The format of the objects is sufficiently standardized to allow straightforward implementation of common operations such as object saving and deletion.

- **Inheritance of features.** It is often the case that several objects differ only in details. For example, two lilac inflorescences may differ only in the color of their petals. The mechanism of inheritance is employed to store such objects efficiently.

- **Version control.** Interaction with an object during experimentation may result in a temporary or permanent modification. In the latter case, the user is able to decide whether the newly created object replaces the old one or should be stored as another version of the original object.

- **Interactive manipulation of objects.** The laboratory provides a set of general-purpose tools for manipulating object parameters. For example, objects can be modified using control panels or by editing specific fields in a textual description of an experiment.

- **Flexibility in conducting experiments.** The user may apply tools to objects in a dynamic way while an experiment is being conducted. This can be contrasted to a static experiment designed when the object is initially incorporated into the system.

- **Guidance through the laboratory.** A hypertext system imposes a logical organization on the set of objects, provides a textual description of the experiments, and makes it possible to browse through the experiments in many ways. Specific experiments are invoked automatically when the corresponding text is selected, in order to facilitate demonstrations and assist a novice user.

So far, objects have been referred to in an intuitive way, relying on the analogy between a real and virtual laboratory. For example, if our interest is in the development of the gametophyte *Microsorium linguaeforme*, in a real laboratory we would experiment with a specimen of the plant, while in a virtual laboratory we explore the corresponding mathematical model. However, the analogy to real objects does not extend to the level of detailed object definition. Specific design decisions are needed for software development purposes. In the current design, a laboratory object is defined as a directory containing two types of files and a subdirectory.

Objects

- The *data files* comprise our knowledge of a particular model.

- A *specification file* defines the data files which make up the object and the tools which apply to them.

- A directory of *extensions* lists objects which inherit some features of the current object.

The object-oriented file structure which provides the basis for lab operation can be represented by a hierarchy of directories and files (Figure A.1).

The path of subdirectories leading to an object establishes the inheritance structure for the lab. Inheritance is based on the idea of specifying new objects in reference to objects which already exist [81]. The "old" object is called a *prototype* and the new one is its *extension*. The extension contains only those files which are different from the corresponding files in the prototype. Files that remain the same are *delegated* to the prototype by establishing links. In other words, the object directory will contain those files that are unique to the object, and links to files that are inherited from its prototype (Figure A.2). This approach saves space, facilitates creation of objects similar to the prototype, and allows a single change in the prototype to propagate through all descendents.

Inheritance of features

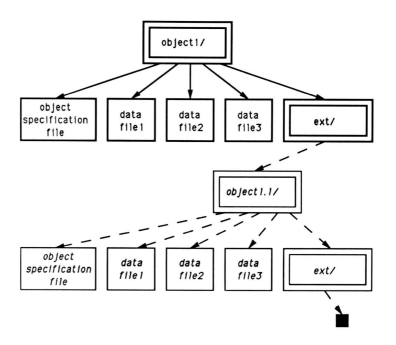

Figure A.1: The hierarchical structure of objects

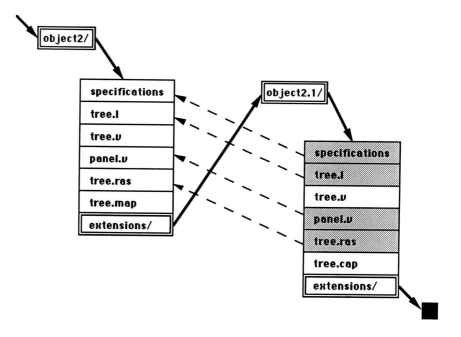

Figure A.2: A prototype and its extension. Shaded areas indicate links.

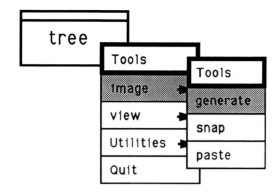

Figure A.3: An object icon with menus

To conduct an experiment, all files that make up the selected object are copied to a temporary location called the *lab table*. Consequently, manipulation of object parameters does not disturb the stored version. When the experiment is finished, the user may save the results by overwriting the original object or by creating an extension. In the latter case, the files on the lab table are compared with those in the prototype object; those files that differ from the prototype are saved, and links to the remaining files are established automatically.

Version control

The ability to manipulate the parameters in an experiment easily is an essential feature of the virtual laboratory. As a rule, all parameters involved in an experiment are supplied to the tools through the object's data files. In order to modify a parameter, the user edits the appropriate file, which is subsequently re-read by the application. Though the editing of parameters can be accomplished using a text editor, in many cases parameter modification can be performed more conveniently using *virtual control panels* [114]. The current implementation of the laboratory provides the user with a general-purpose *control manager* which creates panels according to user-supplied configuration files.

Object manipulation

The user is able to apply a tool to an object as a whole, without detailed knowledge of the programs involved or the component files. This is achieved through the object's specification file which lists all files associated with an object and the tools that can be applied to them. This information is used to create a hierarchy of menus associated with an icon representing the object (Figure A.3). The end nodes in the hierarchy invoke tools that operate on the object. For example, selection of the item **image** followed by the item **generate** from the menus in the figure would invoke the plant modeling program *Pfg*.

Tool application

A user may browse through the objects in the lab by following either the hierarchical structure of objects or hypertext links. The *browser* is used to navigate through the hierarchy, moving down through successive extensions or up through previous levels. At any time, the user may request that an object be placed on the lab table. The hypertext document associated with the lab provides an alternative method of browsing and a means of relating objects independent of the hierarchy.

Browsing

A.2   List of laboratory programs

The essential programs incorporated into the virtual laboratory in botany
are listed below.

- *Plant and fractal generator (Pfg)*
  P. Prusinkiewicz and J. Hanan
  Given an L-system, a set of viewing parameters and optional files
  specifying predefined surfaces, *Pfg* generates the modeled structure by carrying out the derivation, then interpreting the resulting
  string using turtle geometry. Both non-parametric and parametric L-systems are supported. The model can be visualized directly
  on the screen of an IRIS workstation or output to a file. The first
  mode of operation is used to experiment with the model interactively and present developmental sequences. The output file can
  be either in Postscript format, particularly suitable for printing
  results such as fractal curves and inflorescence diagrams on a laser
  printer, or in the format required by the ray-tracer *Rayshade* for
  realistic rendering of the modeled structures.

- *Modeling program for phyllotactic patterns (Spiral)*
  D. R. Fowler
  Spiral is an interactive program for modeling organs with spiral
  phyllotactic patterns. The user can choose between planar and
  cylindrical patterns, and modify parameters which define model
  geometry (Chapter 4). This technique is faster than "growing"
  organs using parametric L-systems. Once an organ has been designed, it can be expressed using an L-system and incorporated
  into a plant structure.

- *Interactive surface editor (Ise)*
  J. Hanan
  Ise makes it possible to define and modify bicubic surfaces consisting of one or several arbitrarily connected patches. The output
  files produced by *Ise* are compatible with *Pfg* and *Spiral*.

- *Modeling program for cellular structures (Mapl)*
  F. D. Fracchia
  Mapl accepts the specification of a two-dimensional cell layer captured by a map L-system and generates the resulting developmental sequence using the dynamic method of map interpretation.
  Options include map generation on the surface of a sphere, and
  the simulation of development in three dimensions according to
  a given cellwork L-system. As in the case of *Pfg*, the models can
  be visualized directly on the screen or output to a file in either
  Postscript or *Rayshade* format.

- *Control panel manager (Panel)*
  L. Mercer and A. Snider
  This program creates control panels containing sliders and buttons, according to a configuration file provided by the user. Upon activation of a control by the mouse, *Panel* generates a message which indicates the corresponding control value. Application programs process this information and modify the appropriate parameters. For example, a panel can be used to control parameters used by *Pfg, Spiral,* or *Mapl.*

- *Ray tracer (Rayshade)*
  C. Kolb, Yale University

  Rayshade reads a scene description from a text file, and renders it using ray tracing. Scenes can be composed of primitives such as planes, triangles, polygons, spheres, cylinders, cones and height fields, grouped together to form objects. These objects can be instantiated in other object definitions to create a hierarchical description of a scene. Transformations including translation, rotation and scaling, and a variety of procedural textures can be applied to any object. Extended light sources, simulation of depth of field, and adaptive supersampling are supported. The program uses 3D grids to partition object space for fast intersection tests.

- *Previewer for the ray tracer (Preray)*
  A. Snider
  Preray is a previewer for *Rayshade* used to provide a fast wire frame rendering of a scene before committing time to ray tracing. A control panel associated with *Preray* makes it possible to set viewing parameters interactively.

- *Modeling program based on Euclidean constructions (L.E.G.O.)*
  N. Fuller
  L.E.G.O. makes it possible to model two- and three-dimensional objects using geometric constructions. In the scope of this book, *L.E.G.O.* was used to model man-made objects such as the *Zinnia* vase and the *Water-lilies* bridge.

- *Iterated function system generator (Ifsg)*
  D. Hepting
  A fractal defined by an iterated function system is described by a finite set of contractive affine transformations with an optional finite state control mechanism. *Ifsg* accepts input from a file specifying the transformations and rendering information. The program is capable of rendering by either attracting, distance-based or escape-time methods. The output can be displayed directly on an IRIS workstation or written to a file for further processing. In the scope of this book, *Ifsg* was used to obtain results which related plant models expressed using L-systems to fractals.

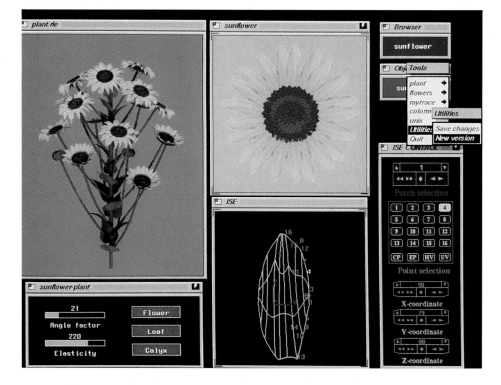

Figure A.4: A virtual laboratory screen

Figure A.4 presents a sample screen of a Silicon Graphics IRIS 4D/60 workstation running some of the above programs within the virtual laboratory framework. The icon in the top right corner represents the laboratory browser which was used to select a sunflower plant as the current object. The icon underneath and the associated menu were subsequently applied to select tools which operate on the object. The control panel in the bottom right corner of the screen is a part of the surface editor *Ise*. The manipulated petal is displayed as a wire frame in the window labeled *Ise*, and incorporated into a flower head by the modeling program *Spiral* which presents its output in the window *sunflower*. The flower heads are in turn incorporated into a complete plant model generated by *Pfg* and rendered using *Rayshade* in the window *plant.rle*. The panel below that window makes it possible to choose organs included in the model and change parameters related to the angles of the branching structure. The metaphor of a virtual laboratory provides a uniform interface to various operations on the selected plant, ranging from the modification of a petal to the rendering of the complete model.

Appendix B

About the figures

The following descriptions of the color images include details about the pictures not described in the main text. Unless otherwise stated, figures were created at the University of Regina.

Figure 1.19 [page 20] *Three-dimensional Hilbert curve*
F. D. Fracchia, P. Prusinkiewicz, N. Fuller
(1989)

This image was rendered using ray-tracing without shadows.

Figure 1.25 [page 26] *Three-dimensional bush*
P. Prusinkiewicz (1986)

Simple branching structure, rendered using the firmware of a Silicon Graphics IRIS workstation. Total generating and rendering time on IRIS 4D/20: 4 seconds.

Figure 1.28 [page 29] *Flower field*
P. Prusinkiewicz (1986)

The field contains four rows of four plants. The scene was rendered with IRIS firmware, using depth-cueing to assign colors to petals.

Figure 1.35 [page 45] *Developmental stages of* Anabaena catenula
J. Hanan, P. Prusinkiewicz (1989)

Figure 2.1 [page 52] *Organic architecture*
Ned Greene, NYIT (1989)

An array of 300 x 300 x 300 voxel space automata was used to

track a polygonal model of a house. Rendering was performed
using a probabilistic radiosity method. See [54] for a full description.

Figure 2.3 [page 54]   *Acer graphics*
                          Jules Bloomenthal, NYIT (1984)

A model of a maple tree. The basic branching structure was generated recursively. Limbs were modeled as generalized cylinders,
obtained by moving discs of varying radii along spline curves.
Real bark texture was digitized and used as a bump map. Leaf
texture was obtained by digitizing a photograph of a real leaf and
emphasizing the veins using a paint program. See [11] for details.

Figure 2.4 [page 54]   *Forest scene*
                          Bill Reeves, Pixar (1984)

A scene from the film *The Adventures of André and Wally B,*
modeled using particle systems. Shading and shadows were approximated using probabilistic techniques. Visible surfaces were
determined using depth-sorting. See [119] for a full description.

Figure 2.5 [page 55]   *Oil palm tree canopy*
                          CIRAD Modelisation Laboratory (1990)

A developmental model of oil palm trees, modeled using the
method originated by de Reffye and described from the graphics perspective in [30].

Figure 2.10 [page 61]   *Medicine lake*
                           F. K. Musgrave, C. E. Kolb, P. Prusinkiewicz,
                           B. B. Mandelbrot (1988)

A scene combining a fractal terrain model, a tree generated using L-systems, and a rainbow. The rainbow model was derived
from a simulation of refraction with dispersion of light through
an idealized raindrop. Procedural textures were applied to the
mountains, the water surface and a vertical plane modeling the
sky. See [101] for further details.

Figure 2.11 [page 62]   *Surrealistic elevator*
                           A. Snider, P. Prusinkiewicz, N. Fuller (1989)

The elevator was modeled using L.E.G.O. The island is a su-
perquadratic surface. Procedural textures were applied to create
stars in the sky, craters on the moon, colored layers in the rock,
waves in the lake and imperfections in the glass that covers the
elevator.

Figure 3.2 [page 69]   *Crocuses*
                       J. Hanan, D. R. Fowler (1990)

The petals were modeled as Bézier surfaces, with the shapes determined using Ise.

Figure 3.4 [page 72]   *Lily-of-the-valley*
                       P. Prusinkiewicz, J. Hanan (1987)

Figure 3.5 [page 74]   *Development of* Capsella bursa-pastoris
                       P. Prusinkiewicz, A. Lindenmayer (1987)

Figure 3.6 [page 75]   *Apple twig*
                       P. Prusinkiewicz, D. R. Fowler (1990)

This twig model was developed in one spring day, looking at a real twig nearby. This time is indicative for most inflorescence models shown.

Figure 3.11 [page 81]   *A mint*
                       P. Prusinkiewicz (1988)

Figure 3.14 [page 84]   *Development of* Lychnis coronaria
                       P. Prusinkiewicz, J. Hanan (1987)

Figure 3.17 [page 90]   *Development of* Mycelis muralis
                       P. Prusinkiewicz, A. Lindenmayer (1987)

Figure 3.18 [page 91]   *A three-dimensional rendering of the* Mycelis *models*
                       P. Prusinkiewicz, J. Hanan (1987)

All internodes in the model are assumed to have the same length. In reality, the internodes have different lengths, and the structure is less crowded.

Figure 3.19 [page 92]   *Lilac inflorescences*
                       P. Prusinkiewicz, J. Hanan, D. R. Fowler (1990)

Figure 3.21 [page 94]   *The Garden of L*
                              P. Prusinkiewicz, F. D. Fracchia, J. Hanan,
                              D. R. Fowler (1988)

All plants were modeled with L-systems and rendered using the
IRIS firmware. Images corresponding to different viewing planes
(the background lilac twigs, the apple twig and the daisies) were
defocused separately using low-pass filters to simulate the depth
of field, then composited with a focused image of lilac inflorescences. The sky was generated using a fractal algorithm.

Figure 3.23 [page 96]   *Wild carrot*
                              P. Prusinkiewicz (1988)

Figure 4.3 [page 102]   *Close-up of a daisy capitulum*
                              D. R. Fowler (1988)

The petals and florets were modeled as Bézier surfaces.

Figure 4.4 [page 102]   *Domestic sunflower head*
                              D. R. Fowler, P. Prusinkiewicz (1989)

Figure 4.5 [page 105]   *Sunflower field*
                              D. R. Fowler, N. Fuller, J. Hanan, A. Snider
                              (1990)

This image contains approximately 400 plants, each with 15 flowers. A flower has 21 petals and 300 seeds, modeled using 600 triangles and 400 triangles respectively. Counting leaves and buds,
the entire scene contains about 800,000,000 triangles. The image
was ray-traced with adaptive supersampling on a grid of 1024 x
768 pixels using 45 hours of CPU time on a MIPS M-120 computer.

Figure 4.6 [page 106]   *Zinnias*
                              D. R. Fowler, P. Prusinkiewicz, J. Hanan,
                              N. Fuller (1990)

The vase was modeled using L.E.G.O. and rendered with a procedural texture. The scene was illuminated by one extended light
source.

Figure 4.7 [page 106]   *Close-up of zinnias*
D. R. Fowler, P. Prusinkiewicz, A. Snider
(1990)

This scene was rendered using distributed ray-tracing to simulate the depth field.

Figure 4.8 [page 108]   *Water-lily*
D. R. Fowler, J. Hanan (1990)

Figure 4.9 [page 108]   *Lily pond*
D. R. Fowler, J. Hanan, P. Prusinkiewicz,
N. Fuller (1990)

The wavelets on the water surface were obtained using bump-mapping with a procedurally defined texture.

Figure 4.10 [page 109]   *Roses*
D. R. Fowler, J. Hanan, P. Prusinkiewicz
(1990)

Distributed ray-tracing with one extended light source was used to simulate depth of field and create fuzzy shadows.

Figure 4.11 [page 111]   *Parastichies on a cylinder*
D. R. Fowler (1990)

Figure 4.15 [page 116]   *Pineapples*
D. R. Fowler, A. Snider (1990)

The image incorporates a physically-based model of a tablecloth approximated as an array of masses connected by springs and placed in a gravitational field. The scene is illuminated by three extended light sources.

Figure 4.16 [page 117]   *Spruce cones*
D. R. Fowler, J. Hanan (1990)

Figure 4.17 [page 117]   Carex laevigata
J. Hanan, P. Prusinkiewicz (1989)

The entire plant, including the leaves, was modeled using parametric L-systems.

Figure 5.2 [page 121] *Maraldi figure*
Ned Greene, NYIT (1984)

The shapes of leaves, calyxes and petals were defined using a paint program, by interpreting gray levels as height. Painted textures were mapped onto the surfaces of leaves and calyxes. Smooth gradation of color across the petals was obtained by assigning colors to the vertices of the polygon meshes representing flowers, then interpolating colors across polygons using Gouraud shading. The vines were rendered with bump-mapping, using a digitized image of real bark.

Figure 5.3 [page 121] *The fern*
P. Prusinkiewicz (1986)

Figure 5.7 [page 125] *A rose in a vase*
D. R. Fowler, J. Hanan, P. Prusinkiewicz (1990)

Petals and thorns are Bézier surfaces incorporated into a rose model expressed using L-systems. The vase was modeled as a surface of revolution.

Figure 6.3 [page 141] *Development of* Anabaena catenula
P. Prusinkiewicz, F. D. Fracchia (1989)

Each developmental stage is plotted in one scan line.

Figure 7.13 [page 161] *Simulated development of* Microsorium linguaeforme
F. D. Fracchia, P. Prusinkiewicz, M. J. M. de Boer (1989)

Cells are represented as polygons, rendered using the IRIS firmware. The development can be visualized directly on the screen of an IRIS 4D/20 workstation without resorting to single-frame animation techniques.

Figure 7.14 [page 161] *Microphotograph of* Microsorium linguaeforme
M. J. M. de Boer, University of Utrecht

Figure 7.16 [page 163] *Simulated development of* Dryopteris thelypteris
F. D. Fracchia, P. Prusinkiewicz, M. J. M. de Boer (1989)

Figure 7.19 [page 169]   *Developmental sequence of* Patella vulgata
F. D. Fracchia, A. Lindenmayer,
M. J. M. de Boer (1989)

Cells are represented as spheres. Intersections of spheres inside
the modeled embryo are ignored, since they do not affect the
ray-traced images.

Figure 7.20 [page 169]   *An electron microscope image of* Patella
vulgata
W. J. Dictus, University of Utrecht

Figure 8.4 [page 180]   *Fern dune*
P. Prusinkiewicz, D. Hepting (1989)

The shape of the leaf has been captured using a controlled iterated function system. A continuous escape-time function defines
point altitudes, resulting in a surrealistic incorporation of a leaf
into the landscape.

Figure 8.9 [page 186]   *Carrot leaf*
D. Hepting, P. Prusinkiewicz (1989)

The leaf shape has been modeled using a controlled iterated function system. The scene consists of a set of spheres, with the radius
equal to the distance to the leaf. The image was rendered using
ray-tracing.

Figure E.1 [page 191]   *Water-lilies*
D. R. Fowler, J. Hanan, P. Prusinkiewicz,
N. Fuller (1990)

A scene inspired by *Water-lilies pool - Harmony in green* by
Claude Monet (1899). All trees and water-lilies were modeled using L-systems. The willow twigs bend downwards due to a strong
tropism effect, simulating gravity. The bridge was modeled using
L.E.G.O. The sky is a sphere with a procedural texture. The
entire scene was ray-traced, then the resulting image was represented as a set of small circles, with the colors close but not equal
to the average of pixel colors underneath. This last operation was
aimed at creating the appearance of an impressionistic painting.

Figure A.4 [page 200]   *Virtual lab*
L. Mercer, D. R. Fowler (1990)

Turtle interpretation of symbols

Symbol	Interpretation	Page
F	Move forward and draw a line.	7, 46
f	Move forward without drawing a line.	7, 46
+	Turn left.	7, 19, 46
−	Turn right.	7, 19
∧	Pitch up.	19, 46
&	Pitch down.	19, 46
\	Roll left.	19, 46
/	Roll right.	19, 46
|	Turn around.	19, 46
$	Rotate the turtle to vertical.	57
[	Start a branch.	24
]	Complete a branch.	24
{	Start a polygon.	120, 127
G	Move forward and draw a line. Do not record a vertex.	122
.	Record a vertex in the current polygon.	122, 127
}	Complete a polygon.	120, 127
~	Incorporate a predefined surface.	119
!	Decrement the diameter of segments.	26, 57
′	Increment the current color index.	26
%	Cut off the remainder of the branch.	73

Bibliography

[1] H. Abelson and A. A. diSessa. *Turtle geometry.* M.I.T. Press, Cambridge, 1982.

[2] M. Aono and T. L. Kunii. Botanical tree image generation. *IEEE Computer Graphics and Applications*, 4(5):10–34, 1984.

[3] M. J. Apter. *Cybernetics and development.* Pergamon Press, Oxford, 1966. (International Series of Monographs in Pure and Applied Biology/Zoology Division Vol. 29).

[4] W. W. Armstrong. The dynamics of tree linkages with a fixed root link and limited range of rotation. *Actes du Colloque Internationale l'Imaginaire Numérique '86*, pages 16–21, 1986.

[5] J. W. Backus. The syntax and semantics of the proposed international algebraic language of the Zurich ACM-GAMM conference. In *Proc. Intl. Conf. on Information Processing*, pages 125–132. UNESCO, 1959.

[6] B. I. Balinsky. *An introduction to embryology.* W. B. Saunders, Philadelphia, 1970.

[7] M. F. Barnsley. *Fractals everywhere.* Academic Press, San Diego, 1988.

[8] M. F. Barnsley, J. H. Elton, and D. P. Hardin. Recurrent iterated function systems. *Constructive Approximation*, 5:3–31, 1989.

[9] B. A. Barsky. The Beta-spline: A local representation based on shape parameters and fundamental geometric measures. PhD thesis, Department of Computer Science, University of Utah, 1981.

[10] R. Bartels, J. Beatty, and B. Barsky, editors. *An introduction to splines for use in computer graphics and geometric modeling.* Morgan Kaufman, Los Altos, California, 1987.

[11] J. Bloomenthal. Modeling the mighty maple. Proceedings of SIGGRAPH '85 (San Francisco, California, July 22-26, 1985) in *Computer Graphics*, 19, 3 (July 1985), pages 305–311, ACM SIGGRAPH, New York, 1985.

[35] P. Eichhorst and W. J. Savitch. Growth functions of stochastic Lindenmayer systems. *Information and Control*, 45:217–228, 1980.

[36] R. O. Erickson. The geometry of phyllotaxis. In J. E. Dale and F. L. Milthrope, editors, *The growth and functioning of leaves*, pages 53–88. University Press, Cambridge, 1983.

[37] G. Eyrolles. Synthèse d'images figuratives d'arbres par des méthodes combinatoires. PhD thesis, Université de Bordeaux I, 1986.

[38] J. B. Fisher and H. Honda. Computer simulation of branching pattern and geometry in Terminalia (Combretaceae), a tropical tree. *Botanical Gazette*, 138(4):377–384, 1977.

[39] J. B. Fisher and H. Honda. Branch geometry and effective leaf area: A study of Terminalia–branching pattern, Parts I and II. *American Journal of Botany*, 66:633–655, 1979.

[40] J. D. Foley and A. Van Dam. *Fundamentals of interactive computer graphics*. Addison-Wesley, Reading, Massachusetts, 1982.

[41] L. Fox and D. F. Mayers. *Numerical solution of ordinary differential equations*. Chapman and Hall, London, 1987.

[42] F. D. Fracchia, P. Prusinkiewicz, and M. J. M. de Boer. Visualization of the development of multicellular structures. In *Proceedings of Graphics Interface '90*, pages 267–277, 1990.

[43] H. Freeman. On encoding arbitrary geometric configurations. *IRE Trans. Electronic. Computers*, 10:260–268, 1961.

[44] D. Frijters. Mechanisms of developmental integration of *Aster novae-angliae* L. and *Hieracium murorum* L. *Annals of Botany*, 42:561–575, 1978.

[45] D. Frijters. Principles of simulation of inflorescence development. *Annals of Botany*, 42:549–560, 1978.

[46] D. Frijters and A. Lindenmayer. A model for the growth and flowering of *Aster novae-angliae* on the basis of table (1,0)L-systems. In G. Rozenberg and A. Salomaa, editors, *L Systems*, Lecture Notes in Computer Science 15, pages 24–52. Springer-Verlag, Berlin, 1974.

[47] D. Frijters and A. Lindenmayer. Developmental descriptions of branching patterns with paracladial relationships. In A. Lindenmayer and G. Rozenberg, editors, *Automata, languages, development*, pages 57–73. North-Holland, Amsterdam, 1976.

[48] M. Gardner. Mathematical games: An array of problems that can be solved with elementary mathematical techniques. *Scientific American*, 216, 1967. 3:124–129 (March), 4:116–123 (April).

[49] M. Gardner. Mathematical games: The fantastic combinations of John Conway's new solitaire game "life." *Scientific American*, 223(4):120–123, October 1970.

[50] M. Gardner. Mathematical games: On cellular automata, self-reproduction, the Garden of Eden and the game "life." *Scientific American*, 224(2):112–117, February 1971.

[51] M. Gardner. Mathematical games – in which "monster" curves force redefinition of the word "curve." *Scientific American*, 235(6):124–134, December 1976.

[52] S. Ginsburg and H. G. Rice. Two families of languages related to ALGOL. *J. ACM*, 9(3):350–371, 1962.

[53] H. Gravelius. *Flusskunde*. Goschen, Berlin, 1914.

[54] N. Greene. Voxel space automata: Modeling with stochastic growth processes in voxel space. Proceedings of SIGGRAPH '89 (Boston, Mass., July 31-August 4, 1989), in *Computer Graphics* 23,4 (August 1989), pages 175–184, ACM SIGGRAPH, New York, 1989.

[55] B. E. S. Gunning. Microtubules and cytomorphogenesis in a developing organ: The root primordium of *Azolla pinnata*. In O. Kiermayer, editor, *Cytomorphogenesis in plants*, Cell Biology Monographs 8, pages 301–325. Springer-Verlag, Wien, 1981.

[56] A. Habel and H.-J. Kreowski. On context-free graph languages generated by edge replacement. In H. Ehrig, M. Nagl, and G. Rozenberg, editors, *Graph grammars and their application to computer science; Second International Workshop*, Lecture Notes in Computer Science 153, pages 143–158. Springer-Verlag, Berlin, 1983.

[57] A. Habel and H.-J. Kreowski. May we introduce to you: Hyperedge replacement. In H. Ehrig, M. Nagl, G. Rozenberg, and A. Posenfeld, editors, *Graph grammars and their application to computer science; Third International Workshop*, Lecture Notes in Computer Science 291, pages 15–26. Springer-Verlag, Berlin, 1987.

[58] F. Hallé, R. A. A. Oldeman, and P. B. Tomlinson. *Tropical trees and forests: An architectural analysis*. Springer-Verlag, Berlin, 1978.

[59] P. H. Hellendoorn and A. Lindenmayer. Phyllotaxis in *Bryophyllum tubiflorum*: Morphogenetic studies and computer simulations. *Acta Biol. Neerl*, 23(4):473–492, 1974.

[60] D. Hepting, P. Prusinkiewicz, and D. Saupe. Rendering methods for iterated function systems. Manuscript, 1990.

[61] G. Herman, A. Lindenmayer, and G. Rozenberg. Description of developmental languages using recurrence systems. *Mathematical Systems Theory*, 8:316–341, 1975.

[62] G. T. Herman and G. Rozenberg. *Developmental systems and languages*. North-Holland, Amsterdam, 1975.

[63] D. Hilbert. Ueber stetige Abbildung einer Linie auf ein Flächenstück. *Mathematische Annalin.*, 38:459–460, 1891.

[64] P. Hogeweg and B. Hesper. A model study on biomorphological description. *Pattern Recognition*, 6:165–179, 1974.

[65] H. Honda. Description of the form of trees by the parameters of the tree-like body: Effects of the branching angle and the branch length on the shape of the tree-like body. *Journal of Theoretical Biology*, 31:331–338, 1971.

[66] H. Honda and J. B. Fisher. Tree branch angle: Maximizing effective leaf area. *Science*, 199:888–890, 1978.

[67] H. Honda and J. B. Fisher. Ratio of tree branch lengths: The equitable distribution of leaf clusters on branches. *Proceedings of the National Academy of Sciences USA*, 76(8):3875–3879, 1979.

[68] H. Honda, P. B. Tomlinson, and J. B. Fisher. Computer simulation of branch interaction and regulation by unequal flow rates in botanical trees. *American Journal of Botany*, 68:569–585, 1981.

[69] H. Honda, P. B. Tomlinson, and J. B. Fisher. Two geometrical models of branching of botanical trees. *Annals of Botany*, 49:1–11, 1982.

[70] R. E. Horton. Erosioned development of systems and their drainage basins, hydrophysical approach to quantitative morphology. *Bull. Geol. Soc. America*, 56:275–370, 1945.

[71] R. E. Horton. Hypsometric (area-altitude) analysis of erosional topology. *Bull. Geol. Soc. America*, 63:1117–1142, 1952.

[72] R. Hunt. *Plant growth analysis*. Studies in Biology 96. Edward Arnold, London, 1978.

[73] R. Hunt. *Plant growth curves – the functional approach to plant growth analysis*. Edward Arnold, London, 1982.

[74] J. E. Hutchinson. Fractals and self-similarity. *Indiana University Journal of Mathematics*, 30(5):713–747, 1981.

[75] G. Van Iterson. *Mathematische und mikroskopish-anatomische Studien über Blattstellungen*. Gustav Fischer, Jena, 1907.

[76] M. Jaeger. Représentation et simulation de croissance des végétaux. PhD thesis, Université Louis Pasteur de Strasbourg, 1987.

[77] J. M. Janssen and A. Lindenmayer. Models for the control of branch positions and flowering sequences of capitula in *Mycelis muralis* (L.) Dumont (Compositae). *New Phytologist*, 105:191–220, 1987.

[78] R. V. Jean. Mathematical modelling in phyllotaxis: The state of the art. *Mathematical Biosciences*, 64:1–27, 1983.

[79] H. Jürgensen and A. Lindenmayer. Modelling development by OL-systems: Inference algorithms for developmental systems with cell lineages. *Bulletin of Mathematical Biology*, 49(1):93–123, 1987.

[80] A. N. Kolmogorov. Three approaches to the quantitative definition of information. *Int. J. Comp. Math*, 2:157–168, 1968.

[81] H. Lieberman. Using prototypical objects to implement shared behavior in object oriented systems. In *Proceedings of the ACM Conference on Object-Oriented Programming Systems, Languages, and Applications*, pages 214–223, New York, 1986. Association for Computing Machinery.

[82] A. Lindenmayer. Mathematical models for cellular interaction in development, Parts I and II. *Journal of Theoretical Biology*, 18:280–315, 1968.

[83] A. Lindenmayer. Adding continuous components to L-systems. In G. Rozenberg and A. Salomaa, editors, *L Systems*, Lecture Notes in Computer Science 15, pages 53–68. Springer-Verlag, Berlin, 1974.

[84] A. Lindenmayer. Developmental algorithms: Lineage versus interactive control mechanisms. In S. Subtelny and P. B. Green, editors, *Developmental order: Its origin and regulation*, pages 219–245. Alan R. Liss, New York, 1982.

[85] A. Lindenmayer. Models for plant tissue development with cell division orientation regulated by preprophase bands of microtubules. *Differentiation*, 26:1–10, 1984.

[86] A. Lindenmayer. Positional and temporal control mechanisms in inflorescence development. In P. W. Barlow and D. J. Carr, editors, *Positional controls in plant development*. University Press, Cambridge, 1984.

[87] A. Lindenmayer. An introduction to parallel map generating systems. In H. Ehrig, M. Nagl, A. Rosenfeld, and G. Rozenberg, editors, *Graph grammars and their application to computer science; Third International Workshop*, Lecture Notes in Computer Science 291, pages 27–40. Springer-Verlag, Berlin, 1987.

[88] A. Lindenmayer. Models for multicellular development: Characterization, inference and complexity of L-systems. In A. Kelmenová and J. Kelmen, editors, *Trends, techniques and problems in theoretical computer science*, Lecture Notes in Computer Science 281, pages 138–168. Springer-Verlag, Berlin, 1987.

[89] A. Lindenmayer and P. Prusinkiewicz. Developmental models of multicellular organisms: A computer graphics perspective. In C. Langton, editor, *Artificial Life: Proceedings of an Interdisciplinary Workshop on the Synthesis and Simulation of Living Systems held September, 1987, in Los Alamos, New Mexico*, pages 221–249. Addison-Wesley, Redwood City, 1989.

[90] A. Lindenmayer and G. Rozenberg, editors. *Automata, languages, development*. North-Holland, Amsterdam, 1976.

[91] A. Lindenmayer and G. Rozenberg. Parallel generation of maps: Developmental systems for cell layers. In V. Claus, H. Ehrig, and G. Rozenberg, editors, *Graph grammars and their application to computer science; First International Workshop*, Lecture Notes in Computer Science 73, pages 301–316. Springer-Verlag, Berlin, 1979.

[92] J. Lück, A. Lindenmayer, and H. B. Lück. Models for cell tetrads and clones in meristematic cell layers. *Botanical Gazette*, 149:1127–141, 1988.

[93] J. Lück and H. B. Lück. Generation of 3-dimensional plant bodies by double wall map and stereomap systems. In H. Ehrig, M. Nagl, and G. Rozenberg, editors, *Graph Grammars and Their Application to Computer Science; Second International Workshop*, Lecture Notes in Computer Science 153, pages 219–231. Springer-Verlag, Berlin, 1983.

[94] N. Macdonald. *Trees and networks in biological models*. J. Wiley & Sons, New York, 1983.

[95] B. B. Mandelbrot. *The fractal geometry of nature*. W. H. Freeman, San Francisco, 1982.

[96] D. M. McKenna. SquaRecurves, E-tours, eddies and frenzies: Basic families of Peano curves on the square grid. In *Proceedings of the Eugene Strens Memorial Conference on Recreational Mathematics and its History*, 1989. To appear.

[97] H. Meinhardt. *Models of biological pattern formation*. Academic Press, New York, 1982.

[98] L. Mercer, P. Prusinkiewicz, and J. Hanan. The concept and design of a virtual laboratory. In *Proceedings of Graphics Interface '90*, pages 149–155. CIPS, 1990.

[99] G. J. Mitchison and Michael Wilcox. Rules governing cell division in *Anabaena*. *Nature*, 239:110–111, 1972.

[100] D. Müller-Doblies and U. Müller-Doblies. Cautious improvement of a descriptive terminology of inflorescences. *Monocot Newsletter 4*, 1987.

[101] F. K. Musgrave, C. E. Kolb, and R. S. Mace. The synthesis and rendering of eroded fractal terrains. Proceedings of SIGGRAPH '89 (Boston, Mass., July 31-August 4, 1989), in *Computer Graphics* 23,4 (August 1989), pages 41–50, ACM SIGGRAPH, New York, 1989.

[102] A. Nakamura, A. Lindenmayer, and K. Aizawa. Some systems for map generation. In G. Rozenberg and A. Salomaa, editors, *The Book of L*, pages 323–332. Springer-Verlag, Berlin, 1986.

[103] P. Naur et al. Report on the algorithmic language ALGOL 60. *Communications of the ACM*, 3(5):299–314, 1960. Revised in Comm. ACM 6(1):1-17.

[104] T. Nelson. Computer lib and dream machines. Self-published, 1980.

[105] P. Oppenheimer. Real time design and animation of fractal plants and trees. *Computer Graphics*, 20(4):55–64, 1986.

[106] G. Peano. Sur une courbe, qui remplit tout une aire plaine. *Math. Annln.*, 36:157–160, 1890. Translated in G. Peano, *Selected works of Giuseppe Peano*, H. C. Kennedy, editor, pages 143–149, University of Toronto Press, Toronto, 1973.

[107] H. Peitgen and D. Saupe, editors. *The science of fractal images.* Springer-Verlag, New York, 1988.

[108] F. P. Preparata and R. T. Yeh. *Introduction to Discrete Structures.* Addison-Wesley, Reading, Massachusetts, 1973.

[109] P. Prusinkiewicz. Graphical applications of L-systems. In *Proceedings of Graphics Interface '86 — Vision Interface '86*, pages 247–253. CIPS, 1986.

[110] P. Prusinkiewicz. Score generation with L-systems. In *Proceedings of the International Computer Music Conference '86*, pages 455–457, 1986.

[111] P. Prusinkiewicz. Applications of L-systems to computer imagery. In H. Ehrig, M. Nagl, A. Rosenfeld, and G. Rozenberg, editors, *Graph grammars and their application to computer science; Third International Workshop*, pages 534–548. Springer-Verlag, Berlin, 1987. Lecture Notes in Computer Science 291.

[112] P. Prusinkiewicz and J. Hanan. *Lindenmayer systems, fractals, and plants*, volume 79 of *Lecture Notes in Biomathematics.* Springer-Verlag, Berlin, 1989.

[113] P. Prusinkiewicz and J. Hanan. Visualization of botanical structures and processes using parametric L-systems. In D. Thalmann, editor, *Scientific Visualization and Graphics Simulation*, pages 183–201. J. Wiley & Sons, 1990.

[114] P. Prusinkiewicz and K. Krithivasan. Algorithmic generation of South Indian folk art patterns. In *Proceedings of the International Conference on Computer Graphics ICONCG '88*, Singapore, 1988.

[115] P. Prusinkiewicz, K. Krithivasan, and M. G. Vijayanarayana. Application of L-systems to algorithmic generation of South Indian folk art patterns and karnatic music. In R. Narasimhan, editor, *A perspective in theoretical computer science — commemorative volume for Gift Siromoney*, pages 229–247. World Scientific, Singapore, 1989. Series in Computer Science Vol. 16.

[116] P. Prusinkiewicz, A. Lindenmayer, and F. D. Fracchia. Synthesis of space-filling curves on the square grid. To appear in *Proceedings of FRACTAL '90*, the 1st IFIP conference on fractals, Lisbon, Portugal, June 6-8, 1990.

[117] P. Prusinkiewicz, A. Lindenmayer, and J. Hanan. Developmental models of herbaceous plants for computer imagery purposes. Proceedings of SIGGRAPH '88 (Atlanta, Georgia, August 1-5, 1988), in *Computer Graphics* 22,4 (August 1988), pages 141–150, ACM SIGGRAPH, New York, 1988.

[118] P. Prusinkiewicz and G. Sandness. Koch curves as attractors and repellers. *IEEE Computer Graphics and Applications*, 8(6):26–40, 1988.

[119] W. T. Reeves and R. Blau. Approximate and probabilistic algorithms for shading and rendering structured particle systems. Proceedings of SIGGRAPH '85 (San Francisco, California, July 22-26, 1985) in *Computer Graphics*, 19, 3 (July 1985), pages 313–322, ACM SIGGRAPH, New York, 1985.

[120] W. R. Remphrey, B. R. Neal, and T. A. Steeves. The morphology and growth of *Arctostaphylos uva-ursi* (bearberry), parts i and ii. *Canadian Journal of Botany*, 61(9):2430–2458, 1983.

[121] W. R. Remphrey and G. R. Powell. Crown architecture of *Larix laricina* saplings: Quantitative analysis and modelling of (non-sylleptic) order 1 branching in relation to development of the main stem. *Canadian Journal of Botany*, 62(9):1904–1915, 1984.

[122] W. R. Remphrey and G. R. Powell. Crown architecture of *Larix laricina* saplings: Sylleptic branching on the main stem. *Canadian Journal of Botany*, 63(7):1296–1302, 1985.

[123] L. H. Reuter. Rendering and magnification of fractals using interated function systems. PhD thesis, Georgia Institute of Technology, 1987.

[124] J. N. Ridley. Computer simulation of contact pressure in capitula. *Journal of Theoretical Biology*, 95:1–11, 1982.

[125] J. N. Ridley. Packing efficiency in sunflower heads. *Mathematical Biosciences*, 58:129–139, 1982.

[126] D. F. Robinson. A notation for the growth of inflorescences. *New Phytologist*, 103:587–596, 1986.

[127] G. Rozenberg and A. Salomaa. *The mathematical theory of L-systems*. Academic Press, New York, 1980.

[128] A. Salomaa. *Formal languages*. Academic Press, New York, 1973.

[129] F. W. Sears, M. W. Zemansky, and H. D. Young. *College physics*. Addison-Wesley Publ. Co., Reading, 6th edition, 1985.

[130] M. Shebell. Modeling branching plants using attribute L-systems. Master's thesis, Worcester Polytechnic Institute, 1986.

[131] P. L. J. Siero, G. Rozenberg, and A. Lindenmayer. Cell division patterns: Syntactical description and implementation. *Computer Graphics and Image Processing*, 18:329–346, 1982.

[132] W. Sierpiński. Sur une courbe dont tout point est un point de ramification. *Comptes Rendus hebdomadaires des séances de l'Académie des Sciences*, 160:302–305, 1915. Reprinted in W. Sierpiński, *Oeuvres choisies*, S. Hartman et al., editors, pages 99–106, PWN – Éditions Scientifiques de Pologne, Warsaw, 1975.

[133] G. Siromoney and R. Siromoney. Rosenfeld's cycle grammars and kolam. In H. Ehrig, M. Nagl, A. Rosenfeld, and G. Rozenberg, editors, *Graph grammars and their application to computer science; Third International Workshop*, Lecture Notes in Computer Science 291, pages 564–579. Springer-Verlag, Berlin, 1987.

[134] G. Siromoney, R. Siromoney, and T. Robinsin. Kambi kolam and cycle grammars. In R. Narasimhan, editor, *A perspective in theoretical computer science — commemorative volume for Gift Siromoney*, Series in Computer Science Vol. 16, pages 267–300. World Scientific, Singapore, 1989.

[135] R. Siromoney and K. G. Subramanian. Space-filling curves and infinite graphs. In H. Ehrig, M. Nagl, and G. Rozenberg, editors, *Graph grammars and their application to computer science; Second International Workshop*, Lecture Notes in Computer Science 153, pages 380–391. Springer-Verlag, Berlin, 1983.

[136] A. R. Smith. Plants, fractals, and formal languages. Proceedings of SIGGRAPH '84 (Minneapolis, Minnesota, July 22-27, 1984) in *Computer Graphics*, 18, 3 (July 1984), pages 1–10, ACM SIGGRAPH, New York, 1984.

[137] A. R. Smith. About the cover: Reconfigurable machines. *Computer*, 11(7):3–4, 1978.

[138] P. S. Stevens. *Patterns in nature*. Little, Brown and Co., Boston, 1974.

[139] R. J. Stevens, A. F. Lehar, and F. H. Perston. Manipulation and presentation of multidimensional image data using the Peano scan. *IEEE Trans. on Pattern Analysis and Machine Intelligence*, PAMI-5(5):520–526, 1983.

[140] A. L. Szilard. Growth functions of Lindenmayer systems. Technical Report 4, Computer Science Department, University of Western Ontario, 1971.

[141] A. L. Szilard and R. E. Quinton. An interpretation for DOL systems by computer graphics. *The Science Terrapin*, 4:8–13, 1979.

[142] R. Thom. *Structural stability and morphogenesis. An outline of a general theory of models.* Benjamin/Cummings, Reading, Massachusetts, 1975.

[143] d'Arcy Thompson. *On growth and form.* University Press, Cambridge, 1952.

[144] W. Troll. *Die Infloreszenzen*, volume I. Gustav Fischer Verlag, Stuttgart, 1964.

[145] W. Troll. *Die Infloreszenzen*, volume II. Gustav Fischer Verlag, Jena, 1969.

[146] A. Turing. On computable numbers with an application to the Entscheidungsproblem, 1936. *Proc. Lond. Math. Soc.* (ser. 2), 42:230–265, 1936–37, and 43:544–546, 1937.

[147] A. Turing. The chemical basis of morphogenesis. *Philosophical Trans. Roy. Soc. B*, 237(32):5–72, 1952.

[148] W. T. Tutte. *Graph theory.* Addison-Wesley, Reading, Massachusetts, 1982.

[149] S. Ulam. Patterns of growth of figures: Mathematical aspects. In G. Kepes, editor, *Module, Proportion, Symmetry, Rhythm*, pages 64–74. Braziller, New York, 1966.

[150] J. A. M. van den Biggelaar. Development of dorsoventral polarity and mesentoblast determination in *Patella vulgata*. *Journal of Morphology*, 154:157–186, 1977.

[151] A. H. Veen and A. Lindenmayer. Diffusion mechanism for phyllotaxis: Theoretical physico-chemical and computer study. *Plant Physiology*, 60:127–139, 1977.

[152] X. G. Viennot, G. Eyrolles, N. Janey, and D. Arquès. Combinatorial analysis of ramified patterns and computer imagery of trees. Proceedings of SIGGRAPH '89 (Boston, Mass., July 31-August 4, 1989), in *Computer Graphics* 23,4 (August 1989), pages 31–40, ACM SIGGRAPH, New York, 1989.

[153] P. M. B. Vitányi. Development, growth and time. In G. Rozenberg and A. Salomaa, editors, *The Book of L*, pages 431–444. Springer-Verlag, Berlin, 1986.

[154] H. Vogel. A better way to construct the sunflower head. *Mathematical Biosciences*, 44:179–189, 1979.

[155] H. von Koch. Une méthode géométrique élémentaire pour l'étude de certaines questions de la théorie des courbes planes. *Acta mathematica*, 30:145–174, 1905.

[156] J. von Neumann. *Theory of self-reproducing automata.* University of Illinois Press, Urbana, 1966. Edited by A. W. Burks.

[157] F. Weberling. Typology of inflorescences. *J. Linn. Soc. (Bot.),* 59(378):215–222, 1965.

[158] F. Weberling. *Morphologie der Blüten und der Blütenstände.* Verlag Eugen Ulmer, Stuttgart, 1981.

[159] H. Weyl. *Symmetry.* Princeton University Press, Princeton, New Jersey, 1982.

[160] S. Wolfram. Computer software in science and mathematics. *Scientific American,* 251(3):188–203, 1984.

[161] S. Wolfram. Some recent results and questions about cellular automata. In J. Demongeot, E. Goles, and M. Tchuente, editors, *Dynamical systems and cellular automata,* pages 153–167. Academic Press, London, 1985.

[162] T. Yokomori. Stochastic characterizations of EOL languages. *Information and Control,* 45:26–33, 1980.

[163] D. A. Young. On the diffusion theory of phyllotaxis. *Journal of Theoretical Biology,* 71:421–423, 1978.

[164] M. H. Zimmerman and C. L. Brown. *Trees — structure and function.* Springer-Verlag, Berlin, 1971.

Index

virtual laboratory 187
voxel space 51

wall 150, 168
    anticlinal 155
    periclinal 155
    tension 151, 173
water-lily 106
wild carrot 95, 131
word 3
    parametric 41
    timed 135

zinnia 106